LaTeX for
Scientists and Engineers

LaTeX for
Scientists and Engineers

David J. Buerger

McGraw-Hill Publishing Company

New York   St. Louis   San Francisco   Auckland   Bogotá   Caracas
Hamburg   Lisbon   London   Madrid   Mexico   Milan   Montreal
New Delhi   Oklahoma City   Paris   San Juan   São Paulo   Singapore
Sydney   Tokyo   Toronto

Library of Congress Cataloging in Publication Data:
Buerger, David J.

    LaTeX for Scientists and Engineers / David J. Buerger

    215 p.
    Bibliography: p. 2
    Includes index. 1. LaTeX (Computer system) 2. Computerized
    typsetting. I. Title.

 234567890   DOCDOC   9543210

ISBN 0-07-008845-4

Printed and bound by R. R. Donnelley and Sons Company.

This book was prepared with LaTeX, and published from camera-ready pages
prepared by the author on a laser printer.

Following are registered and unregistered trademarks of the companies listed:
Amiga, Commodore Business Machines, Inc.; AT, PS/2, IBM, and PC-DOS, In-
ternational Business Machines Corp.; *dBase III Plus*, Ashton-Tate; DVIPS, Ar-
bortext Corp.; Epson FX and Epson LQ, Epson America, Inc.; *FinalWord*, FW
Corp.; Hercules, Hercules Computer Technology; ImageWriter and Macintosh
Plus, Apple Computer Inc.; Lasergrafix and SmartWriter, QMS Corp.; MS-DOS,
Microsoft Corp.; 1-2-3, Lotus Development Corp.; *Perfect Writer*, Thorn/EMI
Computer Software; PostScript, Adobe Systems, Inc.; *Scribe*, Unilogic/Access
Systems Corp; *Sprint*, Borland International; TeX, American Mathematical So-
ciety; VAX and VMS, Digital Equipment Corp.

For more information about other McGraw-Hill materials, call 1-800-2-MCGRAW
in the United States. In other countries, call your nearest McGraw-Hill office.

For Maggie

Contents

List of Figures

List of Tables

Preface

LaTeX is a text formatting system that lets you typeset documents on personal or mainframe computers without the trouble of mastering complex typesetting machines. LaTeX[1] is well-suited for creating technical documents, books, reports, dissertations, manuals, and articles. It lets you create complex mathematical symbols and formulas, foreign language characters, floating tables and figures, table of contents, indexes, bibliographies, cross-references, and footnotes.

Documents created with LaTeX can be printed on a variety of laser and dot-matrix printers. Documents also can be printed on publishing quality phototypesetters. This capability gives you the opportunity for total control over the production of your document.

The LaTeX document preparation system was developed by Leslie Lamport [6]. LaTeX is a collection of special commands, or "macros," based on Donald E. Knuth's TeX program [4]. The system's typesetting therefore is produced by TeX. Unlike TeX, however, LaTeX lets you concentrate on a document's structure and content, rather than on its detailed formatting.

Some instruction books make LaTeX seem as hard to use as TeX. This book was written with the assumption that most users can productively use LaTeX without an exhaustive knowledge of every technical detail. Although LaTeX *for Scientists and Engineers* does not describe every LaTeX nook and cranny, the book's organization lets you pick and choose information germane to the task at hand, without burdening you with unnecessary details. LaTeX *for Scientists and Engineers* was written to provide a fast and easy way to learn how to produce technical documents with LaTeX.

This book assumes that you have no knowledge of TeX or LaTeX. It outlines the basic steps to create typeset documents with LaTeX. Each chapter presents concise discussions and examples. Exercises at the end

---

[1] Usually pronounced *"lah-*tekh," although some prefer *"lay-*tekh," or *"lay-*tex."

of selected chapters help you gain practical experience. Answers to all exercises are at the end of the book. An assortment of sample documents with illustrations of how they are created are in Appendix B. I assume that readers know how to use a word processor or text editor. All discussions and examples are based on the current version (2.09) of LaTeX.

I wish to thank my friends and associates who generously offered comments and encouragement towards the completion of this book. These include David H. Bailey, Nelson H. F. Beebe, James R. Celoni, S.J., Willis Dair, Ronald Danielson, Ruth Davis, Carl Fussell, Tom Golden, Leslie Lamport, Diane Levy, Maggie McClelland, Steven Nahmias, Oren Patashnik, David Raths, John Reuling, Dennis C. Smolarski, S.J., Paul Soukup, S.J., Randall Westgren, and Michael Zomlefer.

Appreciation also is due to Santa Clara University (Santa Clara, California) for providing computer facilities on which this book's numerous drafts were processed.

— D. J. B.

Cupertino, California
November 1, 1989

LaTeX for
Scientists and Engineers

Technical Word Processing and Typography

Professionals in mathematics, the sciences, engineering, and other related fields are just beginning to exploit computer software to produce technical publications. It seems ironic that despite the availability of a multitude of high-powered word processors and other text editing tools, the production of complex technical documents is often a frustrating experience.

The American Mathematical Society, for example, illustrates no less than 19 separate steps to turn a mathematics paper into a publication [12, pp. 5–6]. Clearly, steps such as dealing with publishers and referees will never be computerized. Phases like document creation and proofreading multiple galley drafts, however, can be substantially shortened or eliminated by using the right computer software.

Most technical writers no longer handwrite or typewrite drafts; the use of word processors or text editors on personal computers and large host computers is the practical norm. Yet popular editing software usually does little to help produce complex mathematical formulas, floating tables and figures, or other explanatory tools of the trade. If one needs to produce camera-ready documents, another time-draining issue often rears its head: that of mastering esoteric typesetting conventions. Some who buy and use "desktop publishing" software occasionally find that they now have two full-time jobs.

1.1 Markup Systems and Technical Publishing

A primary goal of technical writing is to communicate complex, logical ideas. Writers employ a host of techniques to accomplish this goal. On an elementary level, rules of punctuation are followed to clarify meaning: words are separated by spaces, periods mark the end of sentences, quotation marks or indentations offset quotations, extra carriage returns separate paragraphs, and numbers are typed to designate outline sections. These methods of marking up text to make it understandable are required whether one uses a word processor, a typewriter, or writes drafts by hand.

More advanced presentational markup techniques such as line centering, boldfacing or italicizing type, and interjecting large typefaces are typical capabilities of sophisticated word processors and desktop publishing software. Punctuation or presentational steps usually are applied manually. At other times, sections of text are tagged for special treatment. These visual systems are highly touted because they let writers see what they are going to get on paper.

Technical writers, however, may find this latter feature to be a liability. Visual markup systems focus attention on appearance, not content. The process of making the document look just right can take more time than producing the document's content. Furthermore, most scientists, mathematicians, and engineers are not trained document designers. Such lack of design expertise may lead to an ineffective presentation of their ideas.

One area in which bad document design can be especially harmful to the clarity of a technical publication is mathematics. Mathematical typesetting demands much more sophisticated design criteria than simple text. It has a larger alphabet of basic symbols, each of which has unique design characteristics different from standard typefaces. Complex spacing criteria apply to formulas, expressions, theorems, proofs, matrices, and arrays. While most visual markup software doesn't provide a way to create complex mathematical symbols and expressions, those few that do assume that users have complete expert knowledge about how they should be formatted.

Another drawback to visual markup software is lack of portability. Such software frequently runs on a limited number of hardware platforms. If a program works only on an Apple Macintosh, for example, it becomes difficult to co-publish a camera-ready paper or book with another author who uses different software on a Sun Microsystems workstation. Organizations that use diverse hardware and software platforms will find internal cooperative publishing difficult.

In light of these problems, many within the mathematical, scientific, and engineering professions have turned to procedural and descriptive markup systems to produce complex technical publications—systems usually not visual in nature. The display monitor shows only text, plus typewritten commands that control how the document will look after it is processed through a formatting program, and then printed on a laser printer or phototypesetter. Some procedural and descriptive systems provide post-processing software that allows one to preview the results as they will appear on paper. A few systems even let you preview the results while the system is processing the file.

Procedural markup programs such as TEX, and UNIX-based *nroff* and *troff*, are complicated to use in their native state. They require the user to type procedure command codes for virtually all aspects of document production. This includes steps as explicit as defining how much space will appear between lines and paragraphs, how far paragraphs are to be indented, and how individual characters are to be used in a document. Such systems' commands can be as complex as a typesetting system, and thus require considerable effort to master.

Descriptive markup systems also require users to type formatting commands into the body of a document's computer file. Unlike procedural markup systems, however, descriptive systems focus the user's attention on document content rather than appearance by having the user type commands that describe what is being typed, instead of how the text should appear in printed form. Technical writers thereby describe sections of text by what they are: Examples include chapter, section, sub-section, table, displayed math formula, and quotation. Printed formats are based on predefined document styles, each created by a typographic designer. Descriptive markup systems ideally free writers from document design concerns, and let them concentrate on content.

In contrast to visual markup software, descriptive markup systems automatically apply mathematical style conventions. Since descriptive markup documents usually are created from straight text files, they can be easily transferred between different computer hardware platforms, and through electronic mail systems.

For more discussion on the philosophy of markup systems, see [3,5], [6, pp. 5–8], and [8].

1.2   LaTeX: A Descriptive Markup System

LaTeX is a document preparation system developed by Leslie Lamport [6]. LaTeX also is a unique type of descriptive markup system, being a collection of special commands—"macros"—based on Donald E. Knuth's TEX program [4]. The system's typesetting therefore is produced by TEX. LaTeX has been implemented on microcomputers, engineering workstations, minicomputers and mainframes using a variety of operating systems. It works virtually the same on all systems. LaTeX enjoys worldwide usage among scientists, mathematicians, engineers, and other scholars and researchers who require its features. This allows great flexibility in producing and sharing typeset documents.

1.3   The Language of Typography

Specialists in every discipline have unique vocabularies that give precision to technical discussions. Not surprisingly, the field of typography has its own jargon. The language of typography becomes pervasive as one moves from typewriters and character-based print towards true typesetting.

Typography is the art of creating styles and arrangements of typeset material. Typographic terms such as *typeface, font, document style, justified text, leading, kerning*, and *serifs* deal with the nature and design of characters, how characters are spaced, how different types of margins affect a document's layout, and a host of other factors.

A typeface, for example, is a particular style of type such as boldface or italic. A font is a family or set of characters in the same typeface and size. A document style predetermines a complete document's layout.

The glossary at the end of this book defines many terms that you will become familiar with as you learn how to use LaTeX. The glossary also includes definitions for the most common LaTeX commands. The index will show you where to find step-by-step instructions on these commands and features.

1.4   What You Need to Use LaTeX

Detailed instructions on hardware and software required to run LaTeX on your computer are found in a book called the *Local Guide*. The *Local Guide* always comes with commercial versions of LaTeX. Institutions that support this software often write and distribute expanded versions tailored to their environment and user needs. The *Local Guide* covers topics such as: How to install LaTeX on your organization's hardware; how to process an input file (i.e. a text file that contains a document's text and formatting commands) through LaTeX; how to use local printer drivers, document previewers, and special style files. The *Local Guide* also contains a list of all fonts available on your system. Depending on how much TeX and LaTeX are used in your institution, the *Local Guide* can vary in size between a short pamphlet and a small book.

In general, input files can be composed on any computer, so long as the resulting text file can be transferred to a computer that has the required software to process and print it. Because LaTeX is a set of macros based on TEX, it cannot run unless you also have TEX installed on the same computer.

LaTEX and TEX are available on many large host systems. If you are uncertain whether they are available at your institution, contact the computer center for more information. Mainframes and engineering workstations offer four distinct benefits that people who use LaTEX on microcomputers may not enjoy:

- Host software often is free or available for a nominal charge. Once installed, every person with a host account can use it.

- Mainframes usually process LaTEX documents much faster than microcomputers. If you are producing a long book or dissertation, the speed advantage could be critical in processing multiple drafts.

- Any microcomputer user with an asynchronous dialup link or network hookup to a mainframe can share LaTEX files with users who otherwise have incompatible hardware.

- Most large hosts and workstation networks have expensive, high-speed laser printers available for general use.

LaTEX and TEX implementations also are available on several different microcomputers. A hard disk is required because the full software distribution can require eight or more megabytes of disk storage space. Most of this space is required to store bit-mapped font files for previewing and printing documents. In addition, you should allow for a few extra megabytes temporarily required to process large files. LaTEX runs on popular microcomputers, including:

IBM and Compatibles. IBM XTs, ATs, PS/2s, and compatible computers require 512 Kbytes of random-access memory (RAM), but work better with 640 Kbytes. Memory-resident programs should be disabled when using IBM-compatible TEX and LaTEX implementations. An AT or 80386-class computer is preferable to an XT because XTs process LaTEX pages very slowly. An AT processes LaTEX files at a speed of roughly one page every 10 seconds compared to approximately 30 seconds per page on an XT. If you want to use software to preview the final form of your text on a monitor before printing it, you will need either a "Hercules-compatible," EGA, or VGA display card; CGA cards will not work with previewers. Commercial and public domain PC printer drivers are available for Epson and Toshiba-compatible dot-matrix printers, and PostScript, Hewlett-Packard, Cordata, QMS Lasergrafix, and Imagen laser printers.

Apple Macintosh. Apple Macintosh Plus computers, and newer models, require at least one megabyte of RAM plus a hard disk. You may prefer to work with a supplementary full-screen display for easier-to-read previewing. Output drivers are available for PostScript laser printers.

Commodore Amiga. Amiga software requires 512 Kbytes of RAM for TₑX and one megabyte of RAM for LATₑX, plus a hard disk. Output drivers are available for the NEC P series, Epson FX and LQ series, ImageWriter II, PostScript, QMS Kiss, and SmartWriter printers.

1.5   Where to Get More Help

In addition to LATₑX *for Scientists and Engineers*, you should obtain and use a copy of your institution's *Local Guide*. It also is wise to cultivate relationships with people who have LATₑX expertise. Such persons can provide valuable suggestions when you are stumped by a thorny formatting problem. Before consulting these people, however, be sure to first check these two books.

If you don't have access to a local expert, you should consider joining the TₑX User's Group (TUG). You can contact TUG at 653 N. Main St., P.O. Box 9506, Providence, Rhode Island, 02940-9506; telephone (401)751-7760. Internet electronic mail can be sent to tug@math.ams.com. TUG conducts TₑX and LATₑX training seminars, and publishes a quarterly journal called *TUGboat*, edited by Barbara Beeton.

Internet electronic mail users should consider subscribing to TUG's informative news digest called TₑXHaX, published in cooperation with the UNIX TₑX distribution service at the University of Washington, and edited by Pierre MacKay and Tiina Modisett. TₑXHaX is a question-and-answer forum whose participants include TₑX and LATₑX aficionados throughout the world. Subscriptions should be sent to this electronic mail address: texhax-request@cs.washington.edu; submissions should be sent to texhax@cs.washington.edu. This digest is also available on Usenet Netnews under the /comp/text newsgroup.

2

Features Overview

LaTeX offers a rich assortment of tools to help writers clearly communicate ideas. This chapter gives a brief overview of some of LaTeX's features. The features introduced in this chapter are explained and illustrated throughout this book. Those of you who use text formatting systems such as *Scribe*, *nroff*, *troff*, *FinalWord*, *Sprint*, and *Perfect Writer* will find much familiar ground in LaTeX.

2.1 Document Structure

LaTeX is a document processing system. What you write is formatted according to a pre-defined document style. LaTeX has four standard document styles: article, book, report, and letter. The style you choose controls how TeX performs the actual typesetting. By changing the document style from one selection to another, you can automatically transform the layout appearance of your entire document into a new format.

Documents can be divided into traditional units like parts and chapters. You also can use more detailed text subdivisions, like sections and subsections. Based on your use of these sectional tools, a table of contents can be automatically generated. Other commands are used to create the title page, appendices, bibliography, and index.

2.2 Typing Text

Text is entered as you usually do with a word processor or text editor. By following a few simple conventions, LaTeX will transform your text input file into a typeset document.

TeX's automatic typesetting attributes include accurate hyphenation, proportional spacing between words, kerned letter combinations, and ligatures. Kerning is a typesetting technique that adjusts the amount of space between two characters. This distance is based on the width of each character. Some character pairs look better when they are moved closer together; some combinations look better when they are spaced farther apart. Ligatures are letter combinations joined together as one unit—for example, ff, fi, fl, ffi, and ffl.

2.3   Type Styles and Sizes

Typeset documents normally do not have <u>underlined</u> text; instead these words are *emphasized* in italic type. In this example, the italics were created by typing {\em emphasized}. The \em switched type to emphasized mode; the curly braces surrounding the word were used as a grouping technique to restrict emphasis to text only within their boundaries.

Other options are described in Section 5.1 that allow you to use **bold-face**, SMALL CAPS, sans serif, *slanted*, and `typewriter` typefaces. Section 5.2 shows how you can change type sizes from this, to this, to this, to this, to this, to this, to this, to this, to this, to this. Appendix D describes how to change LaTeX's default Computer Modern Roman font to other fonts that come with LaTeX distributions.

2.4   Special Characters

A host of multilingual symbols make it easy to create text in some non-English languages. For example, "Nous sommes prêts à partir pour l'Université" is created by typing

```
''Nous sommes pr\^{e}ts \'{a} partir pour l'Universit\'{e}''
```

These symbol commands are listed in Tables 3.1 and 3.2.

2.5   Formatting Environments

Formatting environments allow you to control the appearance of text. An environment is a section of text upon which a specified formatting feature is applied. The typeface changes illustrated above are one example of environment changes. When curly braces are typed *around* a word with the "emphasis" command (e.g. {\em emphasized}), that command is restricted to text *within* the braces; this is analogous to scoping attributes used in computer programming. Another way to create an environment is with \begin{ } and \end{ } commands, much like programmers use subroutines to transfer program control to libraries of sub-programs. This method lets you use preset environments for different types of lists, centered text, poetry, quotations, flush right and flush left text, math symbols and equations, tabs, figures, and tables.

2.6 Math Symbols and Equations

LaTeX provides powerful capabilities for formatting mathematical symbols, expressions, and equations. Formulas such as $\sum_{n=1}^{34} x_n$ can be placed in a text line, or in a "display" mode:

$$\sum_{n=1}^{34} x_n \tag{2.1}$$

Displayed formulas can have automatically generated equation numbers appear in the margin. Another math feature is the capability to produce arrays such as this:

$$\begin{vmatrix} x & 14 & e \\ y-1 & b & f \\ z & c & \lambda \end{vmatrix}$$

Here's an example of a multiline equation:

$$\begin{aligned} |\Pi_n(z_0)|^2 &= |\Sigma_{i=0}^n c_i \phi_i(z_0)|^2 \\ &= |\Sigma_{i=0}^n [c_i \sqrt{N_i}] \cdot [\frac{\phi_i(z_0)}{\sqrt{N_i}}]|^2 \end{aligned}$$

In addition to its equation capabilities, LaTeX has a large assortment of math symbols, including Greek alphabet, binary operation symbols, relation symbols, arrow symbols, log-like function names, and equation delimiters. These are explained and illustrated in Chapter 6, Chapter 7, and in Appendix C.

2.7 Other Document Production Tools

Chapter 8 details how to create figures and tables like Table 2.1. Also useful are footnotes[1] and cross-references, which are discussed in Chapter 9. The

---

[1] Footnotes will wrap to the following page if necessary.

latter can be either forward or reverse references to chapters, sections, tables, figures, equations, and even page numbers. Chapter 11 explains how to create a bibliography, and Chapter 12 discusses the creation of a glossary and index.

Table 2.1. U. S. Paper Currency

Denomination	Picture
$ 1.00	George Washington
$ 5.00	Abraham Lincoln .
$ 10.00	Alexander Hamilton
$ 20.00	Andrew Jackson
$ 50.00	Ulysses Grant
$100.00	Benjamin Franklin

LaTeX provides several style options to be used in conjunction with standard style files. One of the most widely used options is the proceedings style, specially tailored for ACM and IEEE conference proceedings. Many editors now accept LaTeX-produced documents for direct inclusion in published conference proceedings.

Finally, LaTeX's picture environment allows the creation of line drawings within documents. Compared with many graphics software programs now available on personal computers, the picture facility is time-consuming and difficult to use. It will, however, work with any standard TeX printer driver. People with PostScript laser printers may find it easier to create graphics with separate programs and merge the resulting images into LaTeX documents with the \special command. The inclusion of graphics is discussed in Section 14.4.

Beginning Concepts

First-time LaTeX users may find it difficult to understand the general process of creating a document produced with LaTeX. The purpose of this chapter is to take you step by step through this process. This book is a tutorial, and thus includes exercises at the end of many chapters. Answers to exercises are in Appendix A. Most readers will quickly become adept at producing documents with LaTeX by closely reading this tutorial and working through the exercises. If you forget the definition of a term, or cannot remember specific commands, the glossary and index will quickly get you back on track.

3.1 Input Files

A typeset document produced with LaTeX starts with an *input file* that contains the body of text and LaTeX formatting commands. An input file can be created with any word processor or text editor that saves files in ASCII format. Since many popular word processors save files in a unique format, you should check your documentation to find out how to save files in ASCII format. Some word processors refer to ASCII files as "text files," or "unformatted files." These files must not contain any embedded formatting commands unique to your word processor.

3.2 Document Style

Your input file should be named with a .tex extension. The input file's first line must be a \documentstyle command (shown below), which names one of four optional document styles provided with all standard LaTeX distributions: article, book, report, or letter. The article document style formats the text according to pre-defined guidelines for articles and other short documents. While it should be adequate for most documents, you may wish to use one of the alternate styles. The `report` style is suitable for in-depth technical documents. The `book` style is used for actual books such as this one. The `letter` style is ideal for typesetting letters.

To use the `article` style, for example, you must type the following on the first line of your input file:

```
\documentstyle{article}
```

You can designate any of the three alternate document styles by replacing the word "article" with the appropriate style name. The document will be formatted differently, according to guidelines for that type of document style.

Before typing any text for your document's body, you must type one more command after the \documentstyle command:

```
\begin{document}
```

After the end of the body, the last line in your input file must be:

```
\end{document}
```

When the input file is processed through LaTeX, these three commands give the TeX formatter general guidelines for the document's overall appearance, tell it when the actual text body begins, and when it ends. The absence of any of these three commands will produce an error message when the input file is processed with LaTeX.

To prepare your first LaTeX input file, type the following text (cited from [9, p. 33]) and save it into an ASCII file called `SPACE.TEX`:

```
\documentstyle{article}
\begin{document}

The earliest, and still the most fundamental, astronomical
distance determination involves triangulation, the same
method used by surveyors to determine the distance to an
inaccessible point.  The astronomer observes the star of
interest from two different, widely separated points, and
notes the apparent motion of the star against a background
of more distant objects.

\end{document}
```

3.3   How to Run LaTeX

Before you can run LaTeX, you must install both it and TeX on your computer. Consult your *Local Guide* or LaTeX administrator for installation help. Once it is installed, running LaTeX is quite simple. Your first exercise will be to process the sample document listed at the end of Section 3.2 which you saved in an input file called `SPACE.TEX`. To process it, change to the directory that contains `SPACE.TEX` and type:

```
latex space   <return>
```

What you see on the screen after issuing this command varies a bit from one system to the next. If you use LaTeX and TeX on a VAX minicomputer running VMS, for example, you would see something like this:

```
This is TeX, Vax/VMS Version 2.5
(preloaded format=lplain 88.1.19)   15 SEP 1988
**space
(USERS:[JSMITH]SPACE.TEX;1
LaTeX Version 2.09 <15 Sep 1987>
Document Style 'article'. <20 Jul 87>
No file space.aux.
[1] (USERS:[JSMITH]SPACE.AUX;1)
Output written on USERS:[JSMITH]SPACE.DVI;1
(1 pages, 1024 bytes).
Transcript written on USERS:[JSMITH]SPACE.LIS;1
```

Since LaTeX works by virtue of TeX macros, the first things you see announced are the version numbers for each program. Given that we specified the `article` style, this is what LaTeX invokes. The first time you process an input file through LaTeX, an auxiliary (`.aux`) file is created. The `.aux` file is primarily used for cataloging cross-references (discussed in Chapter 9) and table of contents entries (see Chapter 10). It is created regardless of whether you have any such references. As each page is processed you see a page number appear in square brackets. You are then informed how many total pages are in the final document, and how large the file is. If any error messages occur (see Chapter 15), they are recorded in a `.log` or `.lis` file (depending on the system you use).

Note that after you processed `SPACE.TEX` through LaTeX, a new file called `SPACE.DVI` was created. This `.dvi` file (short for *device-independent*) can be examined with a previewer program, or be sent to an output device. Previewer programs let you see the typeset document on your computer screen or terminal as it will appear on paper. Sophisticated previewers let you view documents either one or two pages at a time, and let you shrink or enlarge the image. DVI translation software creates a file that can be sent to your dot-matrix or laser printer, or phototypesetter. Consult your *Local Guide* for instructions on how to install and use these programs on your computer.

The final printed document based on SPACE.DVI will look something like this:

> The earliest, and still the most fundamental, astronomical distance determination involves triangulation, the same method used by surveyors to determine the distance to an inaccessible point. The astronomer observes the star of interest from two different, widely separated points, and notes the apparent motion of the star against a background of more distant objects.

3.4   Special Characters

The process of creating typeset documents requires extra steps beyond what is normally required with word processors, editors, and typewriters. While the bulk of your text is created exactly as with a word processor, all formatting is initiated with LaTeX commands. In particular, you need to pay special attention to the way certain characters are typed to ensure proper treatment by LaTeX. These include:

- LaTeX Command Symbols. Most LaTeX commands are initiated with a backslash ('\') followed by a word, abbreviation, or other symbol. This book largely shows you how to use these commands. Certain frequently used commands do not require the backslash. These include:

$	Starts or terminates math mode.
&	Separates columns in a table.
%	Allows non-printing notes to be added.
#	Designates parameters in a macro.
_	Next character or {block} will be a subscript.
{	Delimiters define block of characters to
}	be treated as a unit.

If you want to produce these symbols in your text, rather than initiate the commands, just type a backslash in front of them. For example, a \\$ will produce $, not the math mode. In short, these seven symbols work backwards: no backslash means command, backslash means symbol.

- **Quotation Marks.** Instead of using the quote mark symbol ("), use a pair of open and closed apostrophe symbols. For example, to create typeset quote marks around the phrase, "Please complete the experiment by Thursday," type the following:

  ```
  ''Please complete the experiment by Thursday.''
  ```

  If you forget to do this and type only the quote mark symbol, you will produce this: "Please complete the experiment by Thursday." Note that the quote mark symbol only creates a *close* quotation mark.

- **Dashes.** Typeset documents contain three kinds of dashes: short ones that connect compound words, medium length ones called "en-dashes" that connect number ranges (e.g. 1987–88), and long "em-dashes" that connect compound phrases (e.g. "Some time ago—perhaps a month or so—we finished writing the course outline."). En-dashes are created by typing two hyphens (e.g. --); em-dashes are created by typing three hyphens (e.g. ---).

  Do not type an input file text line that ends with a dash, as in this example:

  ```
  ''Some time ago---perhaps a month or so---
  we finished writing the course outline.''
  ```

  This causes an extra space to follow the dash in the typeset output file. To prevent this, remember to always place hyphenated words on the same line of text in the input file, like this:

  ```
  ''Some time ago---perhaps a month or so---we finished
  writing the course outline.''
  ```

  Some publication styles require spaces around dashes. Prior examples would thus look like 1987 – 88 and "Some time ago — perhaps a month or so — we finished writing the course outline." The spacing is created by pressing the space bar where you want the space to occur:

  ```
  1987 -- 88
  ''Some time ago --- perhaps a month or so --- we finished
  writing the course outline.''
  ```

Whether you choose to retain or omit spaces around the three types of dashes, be consistent.

- **Multilingual Symbols.** Accent marks are produced by using one of the commands listed in Table 3.1, followed by the character to be accented enclosed in curly braces.

Table 3.1. Accent Commands.

ò	\`{o}	ó	\'{o}	ô	\^{o}
ö	\"{o}	õ	\~{o}	ō	\={o}
ȯ	\.{o}	ŏ	\u{o}	ǒ	\v{o}
ő	\H{o}	o͡o	\t{oo}	ǫ	\c{o}
ọ	\d{o}	o̲	\b{o}		

For example, to create the typeset French phrase, "Les machines à écrire coûtent cher," you would type the following:

```
``Les machines \'{a} \'{e}crire co\^{u}tent cher.''
```

Punctuation and foreign language symbols with their creation commands are listed in Table 3.2. Mathematical symbols are discussed in Section 6.3, and illustrated in Appendix C.

Table 3.2. Punctuation and Foreign Language Symbols.

¶	\P	å	\aa	Œ	\OE
§	\S	Å	\AA	ß	\ss
†	\dag	æ	\ae	ø	\o
‡	\ddag	Æ	\AE	Ø	\O
£	\pounds	ł	\l	¡	!`
©	\copyright	Ł	\L	¿	?`
		œ	\oe		

An example of how one of these symbols is used is this book's copyright notice. It was created by typing:

```
Copyright \copyright 1989 by McGraw-Hill Publishing Co.
```

Note also that whenever punctuation and foreign language symbols are embedded within a word or text string, you must segregate them with the curly brace delimiters. For example, the word "Ångstrom" is created like this:

```
{\AA}ngstrom
```

3.5 Spacing

The typesetting features described in Section 2.2 that benefit LaTeX users come from TeX, which performs the actual formatting. Conceptually, TeX constructs document pages by assembling the individual characters in your input file into words, sentences, paragraphs, and pages. Consider, for a moment, a classical ransom letter we see in detective films. Usually it is a blank piece of paper covered by pasted-on cut-out letters. TeX assembles documents much in the same way, but with more sophistication.

The smallest building block of a TeXed document is the character. The technical TeX term for this building block is a *box*. Much like a criminal pastes letters together to create a ransom note, TeX pastes characters or boxes together to create a typeset document. TeX does its pasteup job with *glue*—also another "TeXnical" term. TeX is especially good at separating characters, words, sentences, and paragraphs with just the right amount of space to create a typeset appearance. This intercharacter spacing is called *kerning*. Controlled interword spacing is called *proportional spacing*.

The vast majority of TeX's work is automatic—you don't need to give it much thought. Occasionally, however, you may wish to influence how TeX does its pasteup job. You already spend a lot of time doing electronic pasteup with regular word processing documents. For example, if you want extra space between words, you simply press the space bar until the desired effect appears. Extra space between sentences or paragraphs is created with the return key. When you use LaTeX, however, you must use some special commands to overcome its automatic typesetting layout. Following are some ways to do this.

- **Between Words and Sentences.** Extra spaces between words in an input file's sentence do not show up in your final document. This means you could begin a line with a space or leave extra spaces between words as you type your document; these don't show up in the typeset document produced with LaTeX. To deliberately insert an extra space, simply type a backslash, preceded and followed by a space. Two extra spaces require a backslash, space, backslash, space. For example, The extra space between words in "Please   complete the experiment by Thursday," is created by typing:

  ```
  ''Please \ complete \ the \ experiment \ by \ Thursday.''
  ```

  Even more space, as in "Please   complete   the   experiment   by Thursday," is created in a similar manner:

```
``Please \ \ complete \ \ the \ \ experiment \ \
by \ \ Thursday.''
```

You can insert a fixed amount of horizontal space between two characters or words with the \hspace{n} command, where hspace stands for "horizontal space" and n is the specified distance. To arbitrarily separate two words by 15 millimeters such as in   this example, you would type:

```
in \hspace{15mm} this
```

You can use whatever dimension you want. Valid LaTeX dimension units include cm (centimeter), em (printers em), in (inch), mm (millimeter), pc (pica), and pt (printer's point).

Words can be pushed to the left   and right margins with the \hfill command. This fills the line with horizontal space when the input file is processed by TeX. The example was created by typing

```
the left \hfill and right
```

When mixing slanted characters such as the italic type (see Chapter 5) with non-slanted text, it's good practice to add an extra space between styles as follows:

It was a *big* insight.   It was a {\it big\/} insight.

The italic correction command \/ adds just enough space to correct for the slant of the previous letter. It does nothing after a roman letter.

Spaces after a period or question mark normally denote the end of a sentence. TeX therefore inserts extra space to denote the start of a new sentence. If a word *within* a sentence ends with a period (e.g. et. al.), and the letter before the period is lower case, type a \ character immediately after the period to ensure proper spacing.

```
(e.g.\ et.\ al.)
```

When a sentence ends with an upper-case letter, you should insert a \@ sequence just before the period for proper spacing. Otherwise TeX assumes it is not the end of a sentence (such as the space between a middle initial and a last name). Here is an example:

"I liked Part I." `''I liked Part I\@.''`

If you want to keep two words from being split across lines, type ~ (a tilde character) as follows:

```
Chapter~15              Mrs.~Robinson
Franklin~D.~Roosevelt   I.R.S.\ Form~1040
91st~Aero Squadron
```

- **Between Lines and Paragraphs.** You can force a line-break by typing \\. A blank line between text lines will start a new paragraph. More than one blank line will not create extra space between paragraphs. You can create extra vertical space between paragraphs by typing \vspace{10mm}. This example would separate two paragraphs by 10 millimeters of space; other dimensions can be substituted as desired. You should not use this command *within* a paragraph; otherwise strange spacing problems will result. For example, the following text:

```
The earliest, and still the most \\ fundamental,
astronomical distance determination involves
triangulation, the same method used by
surveyors to determine \\ the distance to an
inaccessible point.

\vspace{10mm}

The astronomer observes the star of
interest from two different, widely separated
points, and notes the apparent motion of the star
against a background of more distant objects.
```

will create this:

The earliest, and still the most
fundamental, astronomical distance determination involves triangulation, the same method used by surveyors to determine the distance to an inaccessible point.

The astronomer observes the star of interest from two different, widely separated points, and notes the apparent motion of the star against a background of more distant objects.

3.6   In-Text Design Notes

LaTeX provides you with the means to make notes to yourself in the input
file, but it will not print them in the final document. This is called "commenting" your input file. It is done through use of the % character. For
example, this input file text:

```
% This section describes . . .

During the Nixon administration % get Kissinger quote
% and bibliographic information for this section
we saw a vast increase in . . .
```

will produce the following output:

> During the Nixon administration we saw a vast increase in ...

Anything that appears on the same line after a % sign is "invisible" to
LaTeX as it processes your document. Commenting is particularly useful
to describe the purpose of complex LaTeX commands. The comments will
later remind you what you were doing when the input file was created.
If you share your input files with other people (such as in a co-authoring
situation), your comments will inform them what you are doing.

3.7   Summary

The information in this chapter constitutes most of what you need to know
to produce simple LaTeX documents. The rest of this book will instruct you
on techniques required to produce more complex material. For instructions
on printing .dvi files, see your printer driver manual, your *Local Guide*, or
consult your local LaTeX expert.

Problems

Problems in this section, as well as throughout the book, assume that you create an input file with a text editor or word processor, and the appropriate LaTeX commands. It is also assumed that you then process the input file through LaTeX to reach the desired goal.

Problem 3.1 Create the following (from [11, p. 17]):

Vigorous writing is concise. A sentence should contain no unnecessary words, a paragraph no unnecessary sentences, for the same reason that a drawing should have no unnecessary lines and a machine no unnecessary parts. This requires not that the writer make all his sentences short, or that he avoid all detail and treat his subjects only in outline, but that every word tell.

Problem 3.2 Show how the `article` document style is changed to the `report`, `book`, and `letter` options.

Problem 3.3 Create the following text with LaTeX:

The \$ sign is used to start or end the creation of simple mathematical symbols or expressions. The & sign separates columns in tables. Non-printing, in-text notes are preceded by the % sign. Macros often contain designated parameters signified by the # character. Subscripts are created with the _ symbol. When you wish to segregate a block of characters to be treated as one unit, you surround them with the { and } delimiters.

Problem 3.4 Recreate this sentence (from [9, p. 80]). Note the various dash lengths as you type the input.

"Clusters of stars—for example, the Pleiades—are chosen because we can assume that such stellar aggregations are approximately coeval. . . . In Figures 6–5 and 6–6, we see spectrum-luminosity diagrams for two different stellar clusters."

Problem 3.5 Using multilingual symbols, create the following:

Il y eut à toute les époques des enfants qui apprirent à lire, à écrire, à computer; des jeunes gens qui, comme ma sœur, suivirent des cours de lettres et de sciences.

Problem 3.6 A space that follows a period makes TEX think you have reached the end of a sentence. This results in extra space that appropriately denotes such an ending. How do you type a period *within* a sentence and keep TEX from adding the extra space? An example would be "Capt. Roberts."

Problem 3.7 How do you command TEX to insert 5 inches of space between two paragraphs?

Problem 3.8 How do you force a line-break in the middle of a paragraph?

Problem 3.9 Suppose you want to include some document history information (e.g. version number, creation date, author) within an input file, but don't want this information to appear in the printed document. How would you do this?

Formatting Environments

LaTeX works in an *environment-based* system. This means the raw text in your input file is formatted solely by virtue of special commands that you insert into the text. A command retains control over a section, or even an entire document's appearance until it is replaced by another. The entire input file is an environment, because it is subject to the parameters of the \documentstyle named at its beginning. The scope of this command starts with the \begin{document} statement that follows the \documentstyle; it ends with the \end{document} statement at the end of the input file.

Other environments are set up within the boundaries of the overall \documentstyle. These include commands to handle formatting for various types of lists, centered text, poetry, quotations, flush-left and flush-right text, math symbols and equations, tabs, figures, and tables. These environments generally start with a \begin command, and stop with a \end command.

This chapter will explain how to use basic LaTeX environment commands. Advanced environments such as the math mode are discussed in later chapters. While using environments, you should remember two rules: (1) if you \begin an environment, you must \end it at some point; and (2) you can "nest" or combine different environments provided that the nested one is started and stopped before the end of the original environment.

4.1 The "Center" Environment

The center environment is useful for titles. For example, to create

The Quick Brown Fox

by

Millicent the Hen

you would type:

```
\begin{center}
{\it The Quick Brown Fox} \\
by \\
Millicent the Hen
\end{center}
```

If your title has multiple lines, each one except for the last is ended with \\ to force a line break. If you left out the \\ commands you would end up with

The Quick Brown Fox by Millicent the Hen

even if you pressed the return key at the end of each line in your input file. Also, note how the scope of the italic characters in the title is limited by the curly-brace delimiters. The \it command tells the TEX formatter to create italic characters (typeface change details are covered in Chapter 5).

If a paragraph is typed inside the center environment, you will see something like this:

The quick brown fox jumped over the lazy dog. The fox landed in
the lazy dog's food bowl. This made the lazy dog mad. Despite the
lazy dog's anger, however, it didn't bother to punish the quick
brown fox, who zipped down the road to freedom.

4.2   The "Flushleft" and "Flushright" Environments

LATEX's default environment is "fill"—all text fills out to both the left and right margins and is fully justified. The flushleft environment places text flush with the left-hand margin, and leaves text "ragged" on the right. Similarly, the flushright environment places text flush with the right-hand margin, and leaves text "ragged" on the left. For example,

Ms. Susan Johnson
5394 Hamilton Way
San Jose, CA 95126

is produced by this input:

```
\begin{flushright}
Ms. Susan Johnson \\
5394 Hamilton Way \\
San Jose, CA \ 95126
\end{flushright}
```

To produce "ragged-right" text, the flushleft environment should be used. The result will be something like this:

The quick brown fox jumped over the lazy dog. The fox landed in the lazy dog's food bowl. This made the lazy dog mad. Despite the lazy dog's anger, however, it didn't bother to punish the quick brown fox, who zipped down the road to freedom.

4.3 List-Making Environments

You can create three different types of lists with LaTeX:

- Itemized lists (like this one) that preface each item with a bullet.

- Enumerated lists that consecutively (and automatically) number each item. Enumerate commands can be nested to create an outline format.

- Description lists that preface each item with a label.

4.3.1 Itemized Lists

The itemized list in Section 4.3 was created as follows:

```
\begin{itemize}
    \item Itemized lists (like this one) that preface each
        item with a bullet.
    \item Enumerated lists that consecutively number each
        item.  Enumerate commands can be nested to create
        an outline format.
    \item Description lists that preface each item with
        a label.
\end{itemize}
```

Indentations in the input file are not needed; actual input text could appear flush left or in other forms. Note that text for each item is *not* surrounded by curly braces or brackets.

4.3.2 Enumerated Lists

This environment consecutively numbers each item or section of text. If the commands used to create Section 4.3's itemized list format were replaced with \begin{enumerate} and \end{enumerate}, the list would look like this:

1. Itemized lists that preface each item with a bullet.

2. Enumerated lists that consecutively number each item. Enumerate commands can be nested to create an outline format.

3. Description lists that preface each item with a label.

Outlines

By nesting enumerate commands, you can create a complex outline. The following outline was adapted from the table of contents in [12]:

1. Especially for authors
   - (a) The author prepares and submits a manuscript
   - (b) Role of the production editor
2. How to mark mathematical manuscripts
   - (a) Copy editing
   - (b) The copy editor's marks
   - (c) Mathematical expressions
     - i. Fractions
       - A. Stacked fractions
       - B. Numerical fractions
   - (d) Mathematics in display
3. Mathematics in print

This outline was created by typing:

```
\begin{enumerate}
\item Especially for authors
   \begin{enumerate}
   \item The author prepares and submits a manuscript
   \item Role of the production editor
   \end{enumerate}
\item How to mark mathematical manuscripts
   \begin{enumerate}
   \item Copy editing
   \item The copy editor's marks
   \item Mathematical expressions
      \begin{enumerate}
      \item Fractions
         \begin{enumerate}
```

```
    \item Stacked fractions
    \item Numerical fractions
    \end{enumerate}
  \end{enumerate}
 \item Mathematics in display
 \end{enumerate}
\item Mathematics in print
\end{enumerate}
```

Numbering and lettering is automatically created by LaTeX within the **enumerate** environment. Numbering does not appear within bulleted **itemize** lists. LaTeX's default setting can handle nested sections up to four levels deep.

As before, indentations in this sample input are for stylistic illustration. Each input line could have been typed flush left. If you are dealing with a complex outline, however, it may be wise to rely on physical indentations in your input file (as in this example) to make it easier to keep track of when sub-environments begin and end.

4.3.3   Description Lists

This environment is used to create lists of "out-dented" short paragraphs. Each paragraph can have an optional label attached. An example is the following list:

Fox Principal character in the story, *The Quick Brown Fox.* Made a dog mad by jumping over him and landing in his food dish.

Dog Secondary character in the story, *The Quick Brown Fox.* Displayed anger when a fox jumped over him and landed in his food dish. A very lazy character.

Food Dish Used daily by the dog. Underwent possible damage when the fox landed in it.

Note the use of square brackets used to create the optional labels in this list's input file, shown below. Text within the brackets is automatically boldfaced unless you specify a different style command. For example, \item[{\it Fox}] would italicize the word "Fox" instead of printing it in its default boldface type style. Once again, all indentations are for stylistic illustration.

```
    \begin{description}
```

```
\item[Fox] Principal character in the story, {\it
    The Quick Brown Fox}.  Made a dog mad by
    jumping over him and landing in his food dish.

\item[Dog] Secondary character in the story, {\it The
    Quick Brown Fox}.  Displayed anger when a
    fox jumped over him and landed in his food
    dish.  A very lazy character.

\item[Food Dish] Used daily by the dog.  Underwent
    possible damage when the fox landed in it.
\end{description}
```

Custom Description Lists

Sometimes the description environment's default distance between item titles and their descriptions will not be wide enough to make titles stand out. Or you may want descriptions to be fully left-justified for neatness. In either case, you may wish to use a custom description environment such as the one described here.[1]

To create the custom environment called namelist, type the following in your input file's preamble. The preamble is the area between the \documentstyle command and the \begin{document} command):

```
% namelist generates a list with an item width of
% your choice; form: \begin{namelist}{width}
\newcommand{\namelistlabel}[1]{\mbox{#1}\hfil}
\newenvironment{namelist}[1]{%
\begin{list}{}
    {
        \let\makelabel\namelistlabel
        \settowidth{\labelwidth}{#1}
        \setlength{\leftmargin}{1.1\labelwidth}
    }
}{%
\end{list}}
```

This special environment is created using two LaTeX customization commands: \newcommand and \newenvironment. These commands are fully explained in Sections 14.1 and 14.2. To use this environment, you would type the example from Section 4.3.3 as follows:

---

[1] Reproduced courtesy of Nelson H. F. Beebe, Center for Scientific Computing, University of Utah.

```
\begin{namelist}{Food Dishxx}
\item[{\bf Fox}] Principal character in the story,
        {\it The Quick Brown Fox}.  Made a dog mad by
        jumping over him and landing in his food dish.

\item[{\bf Dog}] Secondary character in the story,
        {\it The Quick Brown Fox}.  Displayed anger
        when a fox jumped over him and landed in his
        food dish.  A very lazy character.

\item[{\bf Food Dish}] Used daily by the dog.
        Underwent possible damage when the fox landed
        in it.
\end{namelist}
```

Note that this custom environment is begun with a `\begin{namelist}{ }`
command. The extra pair of curly braces contains the longest item in the
list; the two "x's" are included to create extra space between the items and
descriptions. Unlike the standard `description` environment, the boldface
command for each item (or any other desired attribute) must be added.
The final output looks like this:

Fox   Principal character in the story, *The Quick Brown Fox*. Made a dog
mad by jumping over him and landing in his food dish.

Dog   Secondary character in the story, *The Quick Brown Fox*. Displayed
anger when a fox jumped over him and landed in his food dish. A
very lazy character.

Food Dish Used daily by the dog. Underwent possible damage when the fox
landed in it.

4.4 "Quote" and "Quotation" Environments

Whenever you want to quote something and have its text offset by wider
left and right margins, you should use either the `quote` or `quotation` environment. The difference between them is that the `quote` environment does
not indent the first line of each paragraph, while the `quotation` environment does.

This quotation [11, pp. 71–72],

> Clarity is not the prize in writing, nor is it always the principal
> mark of a good style. There are occasions when obscurity serves
> a literary yearning, if not a literary purpose, and there are writers

whose mien is more overcast than clear. But since writing is com-
munication, clarity can only be a virtue. And although there is no
substitute for merit in writing, clarity comes closest to being one.
...

Clarity, clarity, clarity. When you become hopelessly mired in
a sentence, it is best to start fresh; do not try to fight your way
through against the terrible odds of syntax. Usually what is wrong
is that the construction has become too involved at some point; the
sentence needs to be broken apart and replaced by two or more
shorter sentences.

was created by typing:

```
\begin{quotation}
\small

Clarity is not the prize in writing, nor is it always
the principal mark of a good style.  There are occasions
when obscurity serves a literary yearning, if not a
literary purpose, and there are writers whose mien is
more overcast than clear.  But since writing is
communication, clarity can only be a virtue.  And
although there is no substitute for merit in writing,
clarity comes closest to being one.\ \ldots

Clarity, clarity, clarity.  When you become hopelessly
mired in a sentence, it is best to start fresh; do not
try to fight your way through against the terrible
odds of syntax.  Usually what is wrong is that the
construction has become too involved at some point;
the sentence needs to be broken apart and replaced
by two or more shorter sentences.

\end{quotation}
```

If you like the appearance of smaller type for long quotations, you can use
the \small command as used here (size commands are covered in Chapter 5). So long as the new size command is issued *within* a limited environment, LATEX automatically switches back to the default type size after the
environment is ended. This convention is true within all environments. Be
sure to leave a blank line between the last line of your quotation and the
\end{quotation} command to ensure proper formatting.

4.5   The "Verse" Environment

Poetry is most easily set in the **verse** environment. This poem [1, p. 142],

Gwendolyn Brooks
The Bean Eaters

They eat beans mostly, this old yellow pair.
Dinner is a casual affair.
Plain chipware on a plain and creaking wood,
Tin flatware.

Two who are Mostly Good.
Two who have lived their day,
But keep on putting on their clothes
And putting things away.

And remembering ...
Remembering, with twinklings and twinges,
As they lean over the beans in their rented back room that is full
of beads and receipts and dolls and cloths, tobacco crumbs,
vases and fringes.

1960

was produced by typing:

```
\begin{center}
{\large\bf Gwendolyn Brooks

The Bean Eaters}
\end{center}
\begin{verse}
They eat beans mostly, this old yellow pair. \\
Dinner is a casual affair. \\
Plain chipware on a plain and creaking wood, \\
Tin flatware.

Two who are Mostly Good. \\
Two who have lived their day, \\
But keep on putting on their clothes \\
And putting things away.

And remembering \ldots \\
Remembering, with twinklings and twinges, \\
```

```
As they lean over the beans in their rented back
room that is full of beads and receipts and dolls
and cloths, tobacco crumbs, vases and fringes.

{\footnotesize 1960}
\end{verse}
```

4.6   The "Verbatim" Environment

The verbatim environment reproduces text *exactly* as it appears in your input file. Output is formatted in roman typewriter style, and includes your specified line breaks, spacing, and characters. This means that verbatim output has no proportional spacing, kerning, or ligatures. This environment is useful for things like illustrating computer programming code, or for showing how to type LaTeX input, because backslash commands are treated as ordinary text. The only exception is the \end{verbatim} statement that terminates the environment.

To produce this output:

```
    {\it The Quick Brown Fox\/}
                    is one of
            the
{\it silliest\/} stories I've ever heard.
```

you would type:

```
\begin{verbatim}

    {\it The Quick Brown Fox\/}
                    is one of
            the
{\it silliest\/} stories I've ever heard.

\end{verbatim}
```

If you need to have an \end{verbatim} statement printed, such as in this example, type \verb|\end{verbatim}| *after* the actual \end{verbatim} command. This functions as an "in-text" version of the verbatim environment. Everything inside the delimiter bars is then formatted in the typewriter verbatim mode. Any character can be used to delimit the left and right sides of a \verb command, provided that both characters are identical. For example:

```
\verb#The Quick Brown Fox is a silly story.#
\verb&The Quick Brown Fox is a silly story.&
\verb<The Quick Brown Fox is a silly story.<
\verb+The Quick Brown Fox is a silly story.+
```

The vertical bar character is convenient because it is rarely used in normal text. You can place anything you want inside the delimiter bars. The entire \verb statement, however, must occur on a single line of the input file.

When using the verbatim environment, be sure to type the commands, \begin{verbatim} and \end{verbatim}, flush against the left margin to keep an extra line from appearing in your document.

Problems

Problem 4.1 Create the following with the center environment:

Mathematics Into Type: Copy Editing and Proofreading of Mathematics for Editorial Assistants and Authors
by
Ellen Swanson

Problem 4.2 Create the following flushright example:

Professor Michael Daniels
Department of Economics
231 Sampson Hall
Illinois State University
Chicago, IL 60611

Problem 4.3 Create the following flushleft example:

The earliest, and still the most fundamental, astronomical distance determination involves triangulation, the same method used by surveyors to determine the distance to an inaccessible point. The astronomer observes the star of interest from two different, widely separated points, and notes the apparent motion of the star against a background of more distant objects.

Problem 4.4 Create the following bulletted list of computer network topologies:

- Star
- Hierarchical Tree
- Loop
- Bus
- Ring
- Web

Problem 4.5 Create the following numbered list of computer network topologies:

1. Star
2. Hierarchical Tree
3. Loop
4. Bus
5. Ring
6. Web

Problem 4.6 Create the following list of computer network topologies with their descriptions:

Star: All traffic routed to and handled by a central computer. This topology looks like a spoked wheel, with the central computer at the hub and recipient computers at the end of each spoke.

Hierarchical Tree: Typical to mainframe environments, this topology looks like an upside-down tree. Top computer coordinates network; intermediate computers control traffic at their level and below.

Loop: Typical to workgroups, this topology is a daisy-chain of computers that form a ring. Each computer must be capable of performing all network communications functions.

Bus: A network backbone where all computers share a common communications line (bus). This bus is not joined in a loop, and essentially forms a straight line.

Ring: A cross between a loop and bus topology. Failed nodes, however, do not cause the network to stop working because they attach off the main bus.

Web: A spaghetti-like topology where each node is attached via dedicated links.

Problem 4.7 Create the following quotation (from [4, p. vi]):

Another noteworthy characteristic of this manual is that it doesn't always tell the truth. When certain concepts of TEX are introduced informally, general rules will be stated; afterwards you will find that the rules aren't strictly true. In general, the later chapters contain more reliable information than the earlier ones do. The author feels that this technique of deliberate lying will actually make it easier for you to learn the ideas. Once you understand a simple but false rule, it will not be hard to supplement that rule with its exceptions.

Problem 4.8 Create the following poem (from [1, p. 58]):

H.D. (Hilda Doolittle)
Scribe

Wildly dissimilar
yet actuated by the same fear
the hippopotamus and the wild-deer
hide by the same river.
Strangely disparate

yet compelled by the same hunger,
the cobra and the turtle-dove
meet in the palm-grove.

1920s

Problem 4.9 Recreate this text exactly as you see it here (including the type-writer font).

```
The first command in a \LaTeX\ input file
is this:
```

```
\documentstyle{article}
```

```
An input file's text is surrounded by these commands:
```

```
\begin{document}
```

```
\end{document}
```

Changing the Appearance of Type

Stylistic control is a driving force behind the popularity of typesetting on laser printers and other high-quality output devices. People like to see text not only in book-like fonts, but in different sizes, shapes, and intensities. LaTeX lets you do all of this, and more.

Most word processors create underlines, boldfaces, and other text styles by executing built-in commands. These include clicking on pull-down menu options with a mouse; pressing a function key; or typing a control, shift, or alternate key and specified letter or symbol afterward. The intent is to let you see on the screen what will actually print on a page. Usually this causes unique, non-printable characters to be embedded within the saved, non-ASCII document file.

LaTeX requires that input files be in ASCII format. Its commands are usually initiated by typing a backslash character, followed by a word or symbol that describes what will happen to the text that follows. Since LaTeX commands are written descriptions, input files can be created with any word processor or text editor that saves files in ASCII form.

LaTeX commands are patterned after the *Scribe* text formatting system. If you are familiar with *Scribe* or other systems patterned after it (e.g. *Perfect Writer*, *FinalWord*, *Sprint*), you already know many of LaTeX's formatting requirements.

5.1 Changing Typefaces

LaTeX allows you to easily switch between eight different typefaces. These typefaces and their backslash commands are:

   \rm   Creates roman (normal) type.
   \bf   **Creates boldface type.**
   \it   *Creates italic type.*

\sc   SMALL CAPS TYPE.
\sf   Creates sans serif type.
\sl   *Creates slanted type.*
\tt   `Creates typewriter style type.`
\em   *Creates emphasized type.*

Emphasized type is always italic if the current typeface is roman; if the current typeface is anything other than roman, the emphatic style is roman.

Here is an example of mixing typefaces in a paragraph:

> According to the story, THE QUICK BROWN FOX, the furry featured character jumped *quickly* over the lazy dog. This action made the dog **positively mad.**

One way to create this text is as follows:

```
According to the story, \sc The Quick Brown Fox, \rm the
furry featured character jumped \it quickly\/\rm over
the lazy dog. This action made the dog \bf positively
mad.\rm
```

Note that text following a typeface change command continues in the new style until a \rm command returns text to roman style.

The preferred way to embed typeface commands within text is by grouping them *within* matching pairs of curly braces as follows:

```
According to the story, {\sc The Quick Brown Fox}, the
furry featured character jumped {\it quickly\/} over the
lazy dog.  This action made the dog {\bf positively mad}.
```

This grouping method makes it easier to see where typeface changes start and end. The typeface command is contained within the boundaries set by the curly braces.

Whenever you create a typeface grouping with an open curly brace, you must remember to end it with a matching closed curly brace. If you don't, the typeface you have initiated will continue until a new command is issued. More likely, you'll get a `TeX memory capacity exceeded` error when you process the file through LaTeX.

Note also how the italicized word "*quickly*" is followed by an italic correction (back and forward slashes). This instructs LaTeX to place an extra space between it and the next word. Otherwise the visual effect of "*quickly over*" will not be very attractive. The same technique works with slanted (non-italicized) type. Italic corrections are usually not placed before commas or periods.

5.2 Changing Type Sizes

Type sizes are easily changed with LaTeX. Professional printers usually describe character sizes in terms of a measurement unit called a printer's point. One inch is equivalent to 72.27 points. The point measurement refers to the height of a typeface; LaTeX's default size is 10 points. The term "10 points" is different from "10 pitch," often used by typists to describe a typeface (e.g. Courier) that fits 10 fixed-width characters in one inch of text.

LaTeX's type size commands are similar to typeface backslash commands. They include:

\tiny	Creates type like this.
\scriptsize	Creates type like this.
\footnotesize	Creates type like this.
\small	Creates type like this.
\normalsize	Creates type like this.
\large	Creates type like this.
\Large	Creates type like this.
\LARGE	Creates type like this.
\huge	Creates type like this.
\Huge	Creates type like this.

Table 5.1 shows corresponding point sizes in relation to specified default type sizes.

Table 5.1. LaTeX Type Size Options

Command	10pt Default	11pt Option	12pt Option
\tiny	5pt	6pt	6pt
\scriptsize	7pt	8pt	8pt
\footnotesize	8pt	9pt	10pt
\small	9pt	10pt	11pt
\normalsize	10pt	11pt	12pt
\large	12pt	12pt	14pt
\Large	14pt	14pt	17pt
\LARGE	17pt	17pt	20pt
\huge	20pt	20pt	25pt
\Huge	25pt	25pt	25pt

A document's default type size is set by the \documentstyle command at the beginning of an input file. The command

```
\documentstyle{article}
```

does not specify a type size, so LaTeX assumes that you want to use the default 10-point size. If you want to change the default to 11 points, you would type:

```
\documentstyle[11pt]{article}
```

Once set, the default size prevails throughout the document unless changed by an explicit size command.

5.3   Mixing Typefaces and Sizes

When you invoke a new character size, the typeface switches to roman. You should therefore place typeface commands *after* size commands. For instance, to create

Some **very important advice** is ...

you would type

```
Some {\large\bf very important advice} is \ldots
```

Not every type size is available with each typeface. Table 5.2 lists a matrix of what is available on most standard LaTeX distributions. Fonts marked "P" are preloaded by LaTeX and are available whenever desired. Fonts marked "D" are are only available when you use the \newfont command to demand their use. Appendix D discusses how to use the \newfont command. It also contains printouts of a large variety of fonts and sizes.

Your LaTeX administrator may have added more fonts to your distribution than are indicated on this chart. Check your *Local Guide* for complete information.

5.4   Creating Special Symbols

Occasionally you may want your document to contain special symbols such as the backslash or tilde characters. These symbols normally are used by LaTeX to initiate special commands. To have them appear in your final text, you must create specific symbol definitions.

These definitions may vary depending upon the type of font used to create the symbol. In many cases, a standard ASCII character table will supply the necessary code. Codes for the Computer Modern typewriter font

Table 5.2. Font Sizes Available for LaTeX Typefaces (Font Classes: P = Preloaded, D = Loaded on Demand, X = Unavailable.)

	\it	\bf	\sl	\sf	\sc	\tt
5pt	D	D	X	X	X	X
6pt	X	D	X	X	X	X
7pt	P	D	X	X	X	X
8pt	P	D	D	D	D	D
9pt	P	P	D	D	D	P
10pt	P	P	P	P	D	P
11pt	P	P	P	P	D	P
12pt	P	P	P	P	D	P
14pt	D	P	D	D	D	D
17pt	D	P	D	D	D	D
20pt	D	D	D	D	D	D
25pt	X	D	X	X	X	X

may be different from the Computer Modern italic font. Complete symbol code listings for Donald Knuth's CM fonts are found in [4, Appendices C and F].

For example, the typewriter font's backslash character is created by typing \symbol{'134}. The "134" corresponds to the backslash's octal character code. If a quote mark (") is used in lieu of the apostrophe, you may use the symbol's hexadecimal code.

If your document requires frequent use of a character, you can create special "permanent" definitions with the LaTeX \newcommand to save typing time (full details on this command are in Chapter 14.1). The best place to locate this command is in your input file's preamble (i.e. after the documentstyle command and before the \begin{document} command). For example, to create the backslash definition, type the following:

```
\newcommand{\bs}{\char '134}
```

The \bs sequence is now defined to represent the \ character, as defined by octal character code 134. Once you have defined this character, you can create it anywhere in your document by typing \bs. The reason for creating this definition is to be able to have the backslash appear in your text without having LaTeX think it is a command in your input file. You cannot type \\ to get the \ symbol because \\ by itself is the end-of-line command.

Problems

Problem 5.1 What are two ways to create **boldface** (or for that matter, any alternative to roman) type?

Problem 5.2 Create the following by using LaTeX's 10 type size commands:

This type is tiny.

This type is scriptsize.

This type is footnotesize.

This type is small.

This type is normalsize.

This type is large.

This type is Large.

This type is LARGE.

This type is huge.

This type is Huge.

Problem 5.3 Recreate the following line of text, which combines different type sizes and typefaces (Hint: the new size is **Large**):

Almost every kid knows that bee stings are very painful.

Problem 5.4 Suppose you want to create a ˜ (tilde) character for use in the typewriter mode. How would you do this using the \newcommand procedure? (Hint: the octal character code is 176.)

Simple Math Typesetting

One of LaTeX's big attractions is the flexibility and power it provides in formatting mathematical symbols and equations. People who require these in papers and books formerly have had to improvisionally hand-type them, hoping that typesetters would correctly compose them in the final draft. The purpose of this chapter is to introduce methods of formatting math symbols and simple expressions.

Although the input for examples in this chapter, as well as in Chapter 7, might look a bit daunting at first, it is useful to note that it reads virtually the same as you would orally explain an expression or equation while writing it on a chalkboard. Working through these examples, plus experimenting with some of your own input should get you up to speed in a reasonable period of time.

6.1 Types of Math Environments

LaTeX lets you format math equations three ways. These environments include:

1. math environment that places formulas like $\sum_{n=1}^{100} x_n$ on a line of normal text. The math environment is started by typing a $ sign and ended by the same symbol, or by typing math text between the \(and \) commands, or by using \begin{math} and \end{math}. You should use the $ syntax only for short one-line expressions or equations, mainly because it is easier to forget a $; this makes it hard to find errors when debugging your input file;

2. displaymath environment that offsets formulas in the center of the page following normal text. Here is an example:

$$\sum_{n=1}^{100} x_n$$

The displaymath environment is started by typing either \[, $$,
or \begin{displaymath}. It is ended by typing the corresponding
matching symbols \] or $$, or by \end{displaymath};

3. equation environment, which is identical to displaymath, but adds
an equation number (automatically incremented in each chapter or
article) against the right margin on odd-numbered pages and against
the left margin on even-numbered pages as shown here:

$$\sum_{n=1}^{100} x_n \qquad (6.1)$$

The equation environment is started with \begin{equation} and
ended with \end{equation}.

The fleqn option causes equations in displaymath and equation
environments to be displayed flush against the left margin. This
option can be added to your \documentstyle command as follows:

\documentstyle[12pt,fleqn]{article}

6.2   Simple Expressions and Equations

Math expression and equation formatting is based on the ability to create
subscripts, superscripts, fractions, roots, ellipses, math symbols, and var-
ious log-like functions. Letters inside math mode are italicized and more
widely spaced than in roman type. Numbers in math mode appear as
roman type. Spacing within math formulas is automatically handled by
LATEX. This means that blank spaces in your input document are ignored.
Following are some examples of how to construct basic math expression and
equation components.

• Subscripts and Superscripts are created with the _ and ^ charac-
ters. They can be used alone or combined as seen here:

Output	*Input*
x_2	`$x_2$`
$x_2 y_3$	`$x_{2}y_{3}$`
$y = z_{3b}$	`$y=z_{3b}$`
x^3	`$x^3$`
$x^3 y^2$	`$x^{3}y^{2}$`
5^x	`$5^x$`
$_3 Y^2$	`$_{3}Y^{2}$`
x^{y^3}	`$x^{y^{3}}$`
x^3_2	`$x^{3}_{2}$`
x^3_2	`$x_{2}^{3}$`

- **Fractions** are produced with either the / symbol or the \frac command:

$3/4$	`$3/4$`
$(2/9) * 15.3$	`$(2/9)*15.3$`
$\frac{13}{347}$	`$\frac{13}{347}$`
$x = \frac{y-3}{z}$	`$x = \frac{y-3}{z}$`

Generally the \frac command should be reserved for use in the displaymath or equation environments for best appearances. Following are some examples of complex fractions:

$$x = \frac{y^3 + z/5}{y^2 + 8}$$ `\[ x = \frac{y^{3}+z/5}{y^{2}+8} \]`

$$x = y + \frac{x^3}{z_2} - 4^n$$ `\[ x=y+\frac{x^{3}}{z_{2}}-4^{n} \]`

$$z = \frac{x^{\frac{3}{n-2}}}{y_4}$$ `\[z=\frac{x^{\frac{3}{n-2}}}{y_{4}}\]`

- **Roots** are created with the \sqrt command as shown here:

$\sqrt{2}$	`$\sqrt{2}$`
$\sqrt{n+19}$	`$\sqrt{n+19}$`
$\sqrt[n]{3}$	`$\sqrt[n]{3}$`
$\sqrt{n^3 + \sqrt{y_2}}$	`$\sqrt{n^{3}+\sqrt{y_2}}$`

- Ellipses come in four types: lowered horizontal, centered horizontal, vertical, and diagonal. The lowered horizontal ellipsis (...) is created with the \ldots statement. The \ldots command works outside math mode as well as within. This often is useful when typing normal text. The other three types of ellipses only work in math mode. Vertical and diagonal ellipses are used mainly in arrays (discussed in Chapter 7). Here are some examples:

Type	Output	Input
Lowered	$y_1, \ldots, y_n$	`$y_{1}, \ldots ,y_{n}$`
Centered	$1 + 2 + \cdots + 13$	`$1+2+ \cdots +13$`
Vertical	$\vdots$	`$\vdots$`
Diagonal	$\ddots$	`$\ddots$`

6.3   Math Symbols

Math symbols available with LaTeX can be produced only in math mode. For example, to produce the α character you type `$\alpha$`. LaTeX's symbol commands are illustrated in nine tables found in Appendix C.

If you wish to negate any of these symbols, you may put a slash through them by typing the \not command before it. For example,

If $y^2 \neq 4$ then $y \neq 2$ If `$y^{2}\not= 4$` then `$y\not= 2$`

Usually LaTeX automatically handles spacing intervals between characters, functions, relations, and operators. In certain cases, however, it may need some help. LaTeX provides four horizontal spacing commands that give added control. These sequences include \, (thin space), \: (medium space), \; (thick space), and \! (negative thin space). The latter acts like a backspace command. The thin space command (\,) can be used in any mode; the others must be used in math mode.

Table C.4 (in Appendix C) is a list of "log-like" functions and names that will automatically appear in roman type following a \ when in math mode. For others, simply command roman type, e.g. {\rm sinc}. Here's an example:

$\lim_{n \to \infty} x = 0$ \[\lim_{n \rightarrow \infty} x=0 \]

This same formula in non-display math mode appears as $\lim_{n \to \infty} x = 0$.

Problems

The following math problems are easier to do than they may first appear. Each is based on the foundation of subscripts and superscripts, fractions, roots, and symbols discussed in this chapter. You may wish to practise some of these elements before trying the problems.

Problem 6.1 Create the following expression:

$$\sum_{n=1}^{100} x_n$$

Problem 6.2 Create the following expression:

$$\left| \frac{x^{2k-2}}{\sum_{i=0}^{k-1} a_i x^i} \right|$$

Problem 6.3 Create the following equation:

$$bn \sum_{i=0}^{\log_c n} r^i = bn \frac{r^{1+\log_c n} - 1}{r - 1}$$

Problem 6.4 Create the following statement:

$$\sum_{i=2}^{n-1} i \log_e i \le \int_2^n x \log_e x \, dx \le \frac{n^2 \log_e n}{2} - \frac{n^2}{4}$$

Problem 6.5 Create the following statement:

$$\sum_{i=0}^{n/2-1} (a^2)^i = \prod_{i=o}^{k-2} [1 + (a^2)^{2^i}] = \prod_{i=1}^{k-1} [1 + a^{2^i}]$$

Complex Math Typesetting

7.1 Arrays

The array environment allows you to produce arrays. Each column can be specified to be flush left, centered, or flush right by a one-letter designation (1, c, or r). Each row is separated with the \\ command, except for the last row, where it is omitted. Columns are separated by the & (ampersand) symbol; this symbol must not appear after the last column's contents.

Following is a simple array:

$$
\begin{array}{ccc}
a & 14 & c \\
d-3 & e & f \\
g & h & \lambda
\end{array}
$$

```
\[ \begin{array}{ccc}
a   & 14 & c \\
d-3 & e  & f \\
g   & h  & \lambda
\end{array} \]
```

Here is a more complex array using other LaTeX features:

$$
\det \left|
\begin{array}{lllll}
x_0 & x_1 & x_2 & \cdots & x_n \\
x_1 & x_2 & x_3 & \cdots & x_{n+1} \\
x_2 & x_3 & x_4 & \cdots & x_{n+2} \\
\vdots & \vdots & \vdots & \ddots & \vdots \\
x_n & x_{n+1} & x_{n+2} & \cdots & x_{2n}
\end{array}
\right| > 0
$$

its input being:

```
\[ \det \left|
\begin{array}{lllll}
x_0 & x_1 & x_2 & \cdots & x_n \\
x_1 & x_2 & x_3 & \cdots & x_{n+1} \\
```

```
x_2 & x_3 & x_4 & \cdots & x_{n+2} \\
\vdots & \vdots & \vdots & \ddots & \vdots \\
x_n & x_{n+1} & x_{n+2} & \cdots & x_{2n}
\end{array}
\right| >0 \]
```

Two new features used in this array are the inclusion of items outside the array, and left and right delimiters. As seen in the input example, both of these are typed within the displaymath environment but outside the array environment. Whenever you include delimiters, you must have matching \left and \right statements. You may use different delimiters on each side. Any of the delimiters in Table C.8 can be used. You are not required to actually invoke a delimiter after either of these commands. If you don't want a delimiter on one of the sides, simply type a period (e.g. \left. or \right.).

Delimiters automatically grow to the height of the equation being delimited. You may nest arrays and delimiters as in this example:

$$\left(\begin{array}{c} \left[\begin{array}{cc} a & b \\ c & d \\ e & f \end{array} \right] \\ y \\ z \end{array} \right)$$

```
\[ \left( \begin{array}{c}
\left[ \begin{array}{cc}
a & b \\
c & d \\
e & f
\end{array} \right] \\
y \\
z
\end{array} \right) \]
```

A few other useful array commands are \multicolumn, character alignment, and borders as shown here:

$$\begin{array}{|r@{.}l|} \hline 124 & 34 \\ 45172 & 211 \\ 3 & 8967 \\ \multicolumn{2}{c}{unk.} \\ 792 & 1 \\ \hline \end{array}$$

```
\[ \begin{array}{|r@{.}l|} \hline
124 & 34 \\
45172 & 211 \\
3 & 8967 \\
\multicolumn{2}{c}{\mbox{unk.}} \\
792 & 1 \\
\hline
\end{array} \]
```

This array shows numbers in decimal alignment. Since LaTeX does not provide a true decimal tab feature, we have to improvise. In this example, decimal alignment was made on the period character, designated with the @{.} statement. Any character could be substituted in place of the period. Each | symbol in the \begin{array} statement draws a vertical line at that position (outside or inside the columns). The \hline command draws a horizontal line in each position specified; if it follows a \\ command, a line is drawn under the data on that line. The \multicolumn{ }{ }{ } command allows you to span one entry across several specified columns. The first variable determines how many columns the entry will span across starting from the left column or a given tab marker (&); the second variable instructs how the entry is to be positioned (e.g. centered, left, or right-justified); the third entry is the actual item. The \mbox command, used in this example's displaymath mode, causes all text within its argument to be processed in normal LR text mode.

7.2 Multiline Equations: The "Eqnarray" Environment

Many complex formulas and equations require more than one line. LaTeX provides another environment called eqnarray to meet this need. It works like a three-column array environment, with ampersand characters (&) separating each column. One big difference is that eqnarray does not work in the math mode—it creates its own math mode automatically. Here is a simple multiline equation:

$$
\begin{aligned}
x \ &= \ a - b - c - d \\
&\quad - e - f - g
\end{aligned}
$$

```
\begin{eqnarray*}
x & = & a - b - c - d \\
  &   & \mbox{}- e - f - g
\end{eqnarray*}
```

Note how the \mbox{ } command was used to tell LaTeX to "make an empty text box" (i.e. place a space) before the minus sign preceding the "e". When a formula is split across two lines, the minus sign otherwise would have become a negate sign ("$-e$"; the hyphen is slightly closer to the "e" than a minus sign). To create a negative number or letter (also called a unary operator) within the math environment, type {-e}. If you remove the asterisk in the \begin{eqnarray*} and \end{eqnarray*} commands, each equation automatically becomes consecutively numbered at the right.

Here's another way to break a long formula into two lines:

$$x + y + z + a + b =$$
$$c + d + e + f + g$$

```
\begin{eqnarray*}
\lefteqn{x + y + z + a + b =} \\
&& c + d + e + f + g
\end{eqnarray*}
```

As in the prior example, the two ampersands used in the second line of the equation force it two columns to the right. This convention is used to improve the visual format.

Multiline equations can contain several lines of complex equations. This example,

$$E(X^{-n}, X > t) \approx \frac{\beta^n}{(\alpha - 1)\ldots(\alpha - n)} \int_t^\infty f(x \mid \hat{\mu}, \hat{\sigma}) dx$$

$$= \frac{\beta^n}{(\alpha - 1)\ldots(\alpha - n)} \cdot \left\{ \bar{\Phi}\left(\frac{t - \hat{\mu}}{\hat{\sigma}}\right) \right\}$$

$$= \frac{\mu^n}{\prod_{i=1}^n(\mu^2 - i\sigma^2)} \cdot \left[\bar{\Phi}\left\{ \frac{\mu t - \mu^2 + n\sigma^2}{\sigma\sqrt{\mu^2 - n\sigma^2}} \right\} \right]$$

was created by typing:

```
\begin{eqnarray*}
   E(X^{-n}, X > t) & \approx & \frac{\beta^n}{(\alpha - 1)
      \ldots (\alpha - n)} \int^{\infty}_{t} f(x \mid
      \hat{\mu}, \hat{\sigma}) dx \\
   & = & \frac{\beta^n}{(\alpha - 1) \ldots (\alpha - n)}
      \cdot \left\{\bar{\Phi}\left(\frac{t - \hat{\mu}}
      {\hat{\sigma}} \right)\right\} \\
   & = & \frac{\mu^n}{\prod^{n}_{i = 1} (\mu^{2} - i
      \sigma^{2})} \cdot \left[\bar{\Phi}\left\{\frac{\mu
      t - \mu^{2} + n \sigma^2}{\sigma \sqrt{\mu^{2} - n
      \sigma^{2}}} \right\}\right]
\end{eqnarray*}
```

7.3   Special Effects

Table C.9 lists commands for special accents. Wider hats and tildes can be created with the \widehat{ } and \widetilde{ } commands. Another way to stack symbols is the \stackrel{}{} argument. For example, the symbol $\overset{\rightarrow}{+}$ is created by typing $\stackrel{\rightarrow}{+}$.

You can place lines over or under formulas with the \overline{ } and \underline{ } commands. The latter can be used in normal text mode, although italicized text is preferred to underlining. Examples are:

$\overline{x^2 + y + \overline{z}_3}$

```
$\overline{\overline{x}^{2} + y
   + \overline{z}_{3}}$
```

$2x$ is underlined

```
$\underline{2x}$ is
   \underline{underlined}
```

Braces are created with the \overbrace{ } and \underbrace{ } commands as follows:

$\overbrace{y^2 + y + z} + 3$

```
$\overbrace{y^{2} +
   \underbrace{y + z} + 3}$
```

Problems

Problem 7.1 Create the following matrix:

$$
M = \begin{bmatrix}
0 & 0 & 1 & 2 \\
0 & 0 & 3 & 0 \\
1 & -1 & 0 & 1 \\
2 & 0 & -1 & 3
\end{bmatrix}
$$

Problem 7.2 Create the following Boolean matrix:

$$
\begin{bmatrix}
1 & 0 & 0 & 0 \\
0 & 0 & 1 & 1 \\
1 & 0 & 0 & 1 \\
0 & 0 & 1 & 0
\end{bmatrix}
\begin{bmatrix}
0 & 1 & 0 & 0 \\
1 & 1 & 0 & 0 \\
1 & 0 & 0 & 0 \\
0 & 0 & 0 & 1
\end{bmatrix}
$$

Problem 7.3 Create the following matrix:

i	a_i	x_i	y_i
0	57	1	0
1	33	0	1
2	24	1	-1
3	9	-1	2
4	6	3	-5
5	3	-4	7

Problem 7.4 Create the following parsing table:

	a	b	e	i	t	$
S	$S \rightarrow a$			$S \rightarrow iCtSS'$		
S'			$S' \rightarrow \epsilon$ $S' \rightarrow eS$			$S' \rightarrow \epsilon$
C		$C \rightarrow b$				

Problem 7.5 Create the following equation:

$$T(n) = \begin{cases} O(n), & ifa < c, \\ O(n \log n), & ifa = c, \\ O(n^{\log_c a}), & ifa > c. \end{cases}$$

Problem 7.6 Create the following multiline equation:

$$\begin{aligned} T(m) & \leq & \frac{en}{4m} \left[4M \left(\frac{m}{2} \right) + 4^2 M \left(\frac{m}{2^2} \right) + \cdots + 4^{\log m} M(1) \right] + bnm \\ & \leq & \frac{en}{4m} \sum_{i=1}^{\log m} 4^i M \left(\frac{m}{2^i} \right) + bnm. \end{aligned}$$

Problem 7.7 Create the following sequences using the multiline equation format:

$$\left(\sum_{i=0}^{n-1} a_i x^i \right) \left(\sum_{j=0}^{n-1} b_j x^u \right) = \sum_{k=0}^{2n-2} c_k x^k, \quad \text{where} \quad c_k = \sum_{m=0}^{n-1} a_m b_{k-m}.$$

Tables and Figures

Technical papers and books usually convey lots of detailed information. This material often is best communicated in tabular form, contained in tables or figures. LaTeX provides several ways to position columnar tables of information in text. The simplest method is the **tabbing** environment, which works much like one would use the "tab" key on a typewriter. The more sophisticated **tabular** environment allows you to add lines and boxes around and within a table or figure. Tables and figures also can be made to "float" to an optimal position in the document. This frees you from having to worry about layout issues.

8.1 The "Tabbing" Environment

The **tabbing** environment is most easily understood by looking at a few examples. The following table,

Stock No.	Description	Wholesale Price	Retail Price
24	Disk drive	$55.00	$92.00
32	Monitor	$89.00	$132.00
48	Keyboard	$77.00	$99.00

was produced by typing:

```
\begin{tabbing}
{\it Stock No.} \= {\it Description} \= {\it Wholesale
Price} \= {\it Retail Price} \\
24 \> Disk drive \> \$55.00 \> \$92.00  \\
32 \> Monitor    \> \$89.00 \> \$132.00 \\
48 \> Keyboard   \> \$77.00 \> \$99.00
\end{tabbing}
```

Notice that the "tab stops" were set in the first line with the \= control sequence. Data in the left column were automatically placed flush left. Each of the other columns' data were prefaced by the \> control sequence. The \> LaTeX sequence thus acts like pressing the tab key on a typewriter or word processor; it tells TeX to indent whatever follows it to the appropriate spot. Note that every line except the last ended with the \\ line break command.

There are two alternative ways to produce the same table, each providing you with more control over individual column placement. The first requires that you type spacing characters between each tab setting and end the setup line with the \kill command. The \kill command (shown below) tells LaTeX that there are no more tab stops.

```
\begin{tabbing}
xxxxxxxxxxx\=xxxxxxxxxxxxx\=xxxxxxxxxxxxxxxx\= \kill
{\it Stock No.} \> {\it Description} \> {\it Wholesale
Price} \> {\it Retail Price} \\
24 \> Disk drive \> \$55.00 \> \$92.00  \\
32 \> Monitor    \> \$89.00 \> \$132.00 \\
48 \> Keyboard   \> \$77.00 \> \$99.00
\end{tabbing}
```

Notice how each item in the table's title line now requires a \> sequence to instruct TeX where to place it. When estimating how many spacing characters you will need, be sure to allow enough space for each column's title and contents. If the spacing characters that create tab stops are set too close, such as in this example,

```
\begin{tabbing}
xxxxxx\=xxxxxxxx\=xxxxxx\= \kill
{\it Stock No.} \> {\it Description} \> {\it Wholesale Price}
\> {\it Retail Price} \\
24 \> Disk drive \> \$55.00 \> \$92.00  \\
32 \> Monitor    \> \$89.00 \> \$132.00 \\
48 \> Keyboard   \> \$77.00 \> \$99.00
\end{tabbing}
```

you would end up with something like this:

Stock NDescriptioWholesaleRePaikcPrice
24   Disk drive$55.00 $92.00

```
32      Monitor $89.00  $132.00
48      Keyboard$77.00  $99.00
```

The second method is to include fixed spacing commands as follows:

```
\begin{tabbing}
\hspace{1in}\=\hspace{1.5in}\=\hspace{1.5in}\=
\hspace{1in} \kill
{\it Stock No.} \> {\it Description} \> {\it Wholesale
Price} \> {\it Retail Price} \\
24 \> Disk drive \> \$55.00 \> \$92.00  \\
32 \> Monitor    \> \$89.00 \> \$132.00 \\
48 \> Keyboard   \> \$77.00 \> \$99.00
\end{tabbing}
```

If your table is long or complex in nature, you probably will find the tabular environment easier to use. Nevertheless, there are a few other tabbing command sequences that you may find useful. Command sequences for tabbing are summarized in Table 8.1.

Table 8.1. Tabbing Commands

Command	Where Text is Positioned or Effect
\>	Next tab stop
\<	Prior tab stop
\`	Against right margin
\'	Flush right against prior tab stop
\=	Sets new tab stop at current spot
\\	Starts a new line
\kill	Retains tab stops set on current line without printing its text (used for spacing); starts a new line
\+	Sets new left margin one tab stop to right
\-	Sets new left margin one tab stop to left
\pushtabs	Saves current tab settings within a tabbing environment
\poptabs	Restores settings saved with \pushtabs, within the same tabbing environment

The following example illustrates how some of these tabbing commands could be used:

```
\begin{tabbing}
xxxxxxxxxxxxxx\=                          \kill
No.         \> A-739  \` B-221            \\
1           \> 21                   \+ \\
                 19  \' 56               \\
```

```
\<  2                              \-  \\
3              \> 12    \' not applicable  \\
\pushtabs
Data for 4 not completed                  \\
\poptabs
5              \> 749
\end{tabbing}
```

The resulting table would look like this:

No.	A-739		B-221
1	21		
	19 56		
2			
3	12		not applicable
Data for 4 not completed			
5	749		

As inferred earlier, the tabbing commands needed to produce this table are quite complex. If you require tables that have any degree of complexity, you should first consider using the tabular environment.

8.2 The "Tabular" Environment

The tabular environment allows you to create complex tables and figures, and easily draw boundary lines around and within them. This environment execution is nearly identical to the math array environment. The main difference is that its items are processed in any mode, not just math mode.

This table, showing standard conjugations of the French verb *être*, was created with the tabular environment:

je suis	nous sommes
tu es	vous êtes
il est	ils sont

Here is how this table was created:

```
\begin{center}
\begin{tabular}{|c|c|} \hline
je suis & nous sommes \\ \hline
tu es & vous \^{e}tes \\ \hline
il est & ils sont \\ \hline
\end{tabular}
\end{center}
```

The \begin{center} command horizontally placed the table in the center of the page. Following the \begin{tabular} command, the vertical bars created the vertical lines within the table. The "c" statements centered the contents of each column. We could have substituted the letter "r" or "l" for right or left justification. The \hline command draws a horizontal line in each position specified; if it follows a \\ command, a line is drawn under the data on that line.

LaTeX allows you to add other features to tables, such as columnar titles and variable width columns. This table, containing additional *être* conjugations,

Imparfait	
j'	étais
tu	étais
il	était
nous	étions
vous	étiez
ils	étaient

Plus-que-parfait		
j'	avais	été
tu	avais	été
il	avait	été
nous	avions	été
vous	aviez	été
ils	avaient	été

was created with these commands:

```
\begin{center}
\begin{tabular}{|ll|l|lll|} \cline{1-2} \cline{4-6}
\multicolumn{2}{|c|}{\sl Imparfait} & \hspace{7mm} &
\multicolumn{3}{|c|}{\sl Plus-que-parfait} \\
                                       \cline{1-2} \cline{4-6}
j'   & \'{e}tais   &   & & j'   & avais   & \'{e}t\'{e} \\
tu   & \'{e}tais   &   & & tu   & avais   & \'{e}t\'{e} \\
il   & \'{e}tait   &   & & il   & avait   & \'{e}t\'{e} \\
     &             &   & &      &         &             \\
nous & \'{e}tions  &   & & nous & avions  & \'{e}t\'{e} \\
vous & \'{e}tiez   &   & & vous & aviez   & \'{e}t\'{e} \\
ils  & \'{e}taient &   & & ils  & avaient & \'{e}t\'{e} \\
\cline{1-2} \cline{4-6}
\end{tabular}
\end{center}
```

The blank space between the left and right portions of this table was created with the p{7mm} statement in the \begin{tabular} command's third column option, plus the \hspace{7mm} command on the first row of the table. These commands allows you to specify a fixed column width in any valid LaTeX dimension unit. The \cline{a-b} command draws a horizontal line across the specified columns (in this case, 1–2 and 4–6). The \multicolumn{ } { }{ } command allows you to span one entry across several specified columns. The first variable determines how many columns the entry will span across starting from the given tab marker (&); the second variable instructs how the entry is to be positioned (e.g. centered, left, or right-justified); the third entry is the actual item.

8.3   The "Table" and "Figure" Environments

Tables created with the tabbing or tabular environments—or for that matter, any text or vertical space created in your input file—can be positioned in an optimal spot in a document. This is done by enclosing it in either the table or figure environments. The only difference between these environments is in how the information is labeled (i.e. *Table* 1 vs. *Figure* 1).

The \begin{table}[ ] or \begin{figure}[ ] command will create a table or figure. These environments are ended by a \end{table} or \end{figure} command. Optional positioning commands are placed in the square brackets. These options are listed in Table 8.2.

Table 8.2. Table/Figure Positioning Commands

Command	Table Placement
b	bottom of page
h	current line ("here")
t	top of page
p	separate float page

Optional positioning commands might not always work as intended. LaTeX formats documents on a page-by-page basis. If you specify a "bottom" position for a table on a page where the amount of existing text plus the table height exceeds the page length, the table will be placed at the next optional position. Multiple options are permitted (e.g. [bht]). If no options are specified, a default placement argument of [tbp] is assumed.

Each table or figure is automatically numbered—through the entire document with the article style, and within chapters for the book and report styles. You also can add a caption to the number. Table 8.2's caption was created by adding this command immediate after the \begin{table} sequence:

```
\caption{Table/Figure Positioning Commands}
```

By typing this command in that spot, the caption appeared above the table. If you or your publisher prefer the caption to appear just below the table or figure, type the same command just before the \end{table} or \end{figure} command. Technical publications usually prefer that tables be captioned at the top, with figures at the bottom. The caption placement choice is a matter of style—whichever you chose, be consistent throughout your document.

Problems

Problem 8.1 Create the following table three different ways:

EIA Designation CCITT Designation Name
AA	101	Protective Ground
AB	102	Signal Ground
BA	103	Transmitted Data
BB	104	Received Data
CA	105	Request to Send
CB	106	Clear to Send
CC	107	Data Set Ready
CF	109	Data Channel Received Line Signal Detector

Problem 8.2 Recreate the table in Problem 8.1 using the tabular environment. The result should look like this:

EIA Designation	CCITT Designation	Name
AA	101	Protective Ground
AB	102	Signal Ground
BA	103	Transmitted Data
BB	104	Received Data
CA	105	Request to Send
CB	106	Clear to Send
CC	107	Data Set Ready
CF	109	DC Rec'd Line Signal Detec.

Problem 8.3 Recreate the table in Problem 8.2 using the tabular environment, but enclose the table within the table environment. Add a caption to the top of the table so that it looks like this:

Table 8.3. Low-Speed Asynchronous Full-Duplex Modem Interface Leads

EIA Designation	CCITT Designation	Name
AA	101	Protective Ground
AB	102	Signal Ground
BA	103	Transmitted Data
BB	104	Received Data
CA	105	Request to Send
CB	106	Clear to Send
CC	107	Data Set Ready
CF	109	DC Rec'd Line Signal Detec.

Footnotes and Cross-References

9.1 Footnotes

The author-date method of making in-text references has, in many fields, become the preferred format.[1] In some cases, however, the use of footnotes is appropriate. If you need to create footnotes, LaTeX easily handles the job. LaTeX's default style files do not allow for the production of endnotes, which appear at the end of a document instead of at the bottom of each page.

To create the footnote seen on this page, the following was typed:

```
. . . preferred format.\footnote{See Chapter 15 in {\it The
Chicago Manual of Style}, 13th ed.\  (The Univ.\ of Chicago
Press, 1982).}  In some cases . . .
```

As you can see, the footnote is embedded within the text, prefaced by the \footnote command and surrounded by curly braces. Style changes within the footnote are allowed. Footnote numbers are automatically assigned by the document style. The article style consecutively numbers footnotes from the beginning to the end of the document. The report and book styles consecutively number footnotes within individual chapters. If you decide to go back later and insert or delete footnotes, the footnotes are automatically renumbered.

One thing to remember: when creating footnotes, the footnote number will appear exactly where the \ of the footnote command occurs. This means you should place the \ immediately after the word or punctuation mark where you want the number to occur, with no spaces in between.

---

[1]See Chapter 15 in *The Chicago Manual of Style*, 13th ed. (The Univ. of Chicago Press, 1982).

Otherwise, your final document will have a space before the footnote number.

9.2 Cross-References

LaTeX's section scheme (see Section 10.1) makes it easy to create forward and backward cross-references to chapters, sections, tables, and figures. Other environments that will generate numbers that can be cross-referenced include enumerate (item's number is assigned), eqnarray, equation, and theorem-like environments defined by the \newtheorem command (discussed in Section 14.2).

If you are planning to refer to one of these four latter items by means of a cross-reference, labels can be assigned that allow for automatic cross-referencing. In this manner, you won't have to remember a specific chapter or figure number when typing the cross-reference. If other sectional units are inserted or deleted after your document is created, cross-references will automatically adjust to reflect their new numbers.

To illustrate how one might use cross-references in a document, consider the earlier discussion of changing type sizes in Chapter 5. This occurred in Section 5.1 starting on page 37. Table 5.1 on page 39 provided comparative information on the point sizes of fonts created with different type styles and sizes. The specific numbers referring to sectional units and their corresponding page numbers created elsewhere in this book are cross-references.

To create these cross-references, I first had to create sectional unit labels. Note the label command following Chapter 5's \chapter command:

```
\chapter{Changing Type Styles and Sizes}\label{ch:change}
```

Section 5.1's command was created like this:

```
\section{Changing Type Styles}\label{sec:style}
```

And Table 5.1's command was

```
\caption{Type Sizes for \LaTeX\ size-changing
    commands.}\label{tab:sizes}
```

These labels were all assigned *after* the appropriate unit sectioning command was created. Here's the input paragraph where the cross-references were made:

```
To illustrate how one might use cross-references in a
document, consider the earlier discussion of changing type
```

```
sizes in Chapter~\ref{ch:change}. This occurred in
Section~\ref{sec:style} starting on page~\pageref{sec:style}.
Table~\ref{tab:styles} on page~\pageref{tab:styles} provided
comparative information . . .
```

It is not necessary to use a prefix and colon in label references. The label can use letters, digits, or punctuation characters. I coded the labels with a short preface—"sec" for section, "ch" for chapter, and "tab" for table—to avoid confusion. Sectional cross-references are made with the \ref{ } command. Requests for page numbers on which a reference occurs are made through the \pageref{ } command. Each title (e.g. Chapter, Section, Table) should be followed by a tie or tilde character (~) to prevent line breaking at that point. See page 19 for more discussion of this feature.

Figures and tables must be labeled *after* the \caption command or the labels won't work.

If you refer to a label that occurs later in the document, you will notice a LaTeX warning: reference undefined message as the input file is processed by LaTeX. These labels are stored in the .aux file. To take care of these references, you will need to process your input file through LaTeX one more time. If you don't, these undefined references will be printed as "??". Generally it's a good idea to process input files through LaTeX two times whenever labels and cross-references are used. If cross-reference changes are made to your input file (e.g. page numbers change or new cross-references are added), LaTeX will issue the following warning message: Label(s) may have changed. Rerun to get cross-references right. Reprocess the input file through LaTeX to take care of cross-reference changes.

Organizing a Document

Books, reports, and articles traditionally are organized with a title or title page, table of contents, lists of figures and tables, chapters, parts, sections, sub-sections, sub-sub-sections, paragraphs, sub-paragraphs, sub-sub-paragraphs, appendices, bibliography, and index. All of these conventions can be created with LaTeX to give you beautifully formatted documents.

10.1   Section Commands

Section commands are the way you create organized divisions in your document. LaTeX automatically assigns numbers to each one. Creating sections is easy. This chapter's command was

    \chapter{Organizing a Document}

This section's command was

    \section{Section Commands}

Sub-sections (that naturally lie beneath sections) are created as follows:

    \subsection{This is a Sub-section}

And sub-sub-sections are created by typing

    \subsubsection{This is a Sub-sub-section}

Remember that the major division for the book and **report** document styles is the \chapter, while the \article document style's major division is the \section.

If you don't want sectional divisions to be numbered in the text or to appear in the table of contents, insert an asterisk before the curly brace as follows:

```
\subsection*{The Art of Creating Subsections}
```

The text in a sectional division marker's curly braces is what appears in the table of contents. If you wish to have different text appear in the table of contents, you can type the option in square brackets as follows:

```
\subsection[Creating Subsections]{The Art of Creating Subsections}
```

The \part division can be used to separate major divisions of long documents. Sequential numbering of small units contained within a part (such as a chapter or section) does not print the part number with these units.

10.2   Title and Title Page

The title is produced immediately following the preamble (discussed in Section 10.4). Here is how to create a title:

```
\begin{document}
\title{\bf \LaTeX\ Quick Reference \\
       for \\
       Scientists and Engineers}
\author{\bf David J. Buerger}
\date{\today}
\maketitle
```

Title information is automatically centered. It is composed of three types of information: title, author, and date. If you want to omit author or date information, leave the curly braces empty (e.g. { }). All three commands must be given, even if one or two are left empty. Note that this example's date used the \today command, which entered the date when the document was processed. You may elect to type a specific date (e.g. March 1988) instead of using the \today command. The \maketitle command actually produces the title.

The book and report styles place the title on a separate page. The article style does not, unless you include the following option in your document style command:

```
\documentstyle[11pt,titlepage]{article}
```

You can produce an abstract placed below the title information (not available for the book style) by typing the following command *before* the \maketitle command.

```
\begin{abstract}
The abstract's text is typed here
\end{abstract}
```

10.3   Table of Contents

A table of contents is produced after the title page and before Chapter 1 at the spot where you type

```
\newpage
\pagenumbering{roman}
\tableofcontents
```

The \newpage command forces a page break and allows the table of contents to be placed on a new page. The page number command paginates the table of contents in lower-case roman numerals. By issuing the command, \pagenumbering{arabic} after the chapter command in your first chapter, pagination is returned to arabic numerals starting with page number 1.

Information in the table of contents is based on data collected in the .aux file as LaTeX processes the document. You must run LaTeX twice on input files that contain the \tableofcontents command.

If you want to include a list of figures and/or a list of tables, simply add the commands

```
\listoffigures
\listoftables
```

immediately after the \tableofcontents command.

10.3.1   Preface or Foreword

Including a preface or foreword to a book or report is easy, but you will need to issue a few extra commands to get the formatting right. Usually the idea is to have this section appear as a chapter but not have a chapter number, and to have this properly reflected in the table of contents.

Assuming you are using the roman page numbering just described, type the following commands right after \tableofcontents to create a preface:

```
\chapter*{Preface}
\addcontentsline{toc}{chapter}{Preface}
\pagestyle{myheadings}
\markboth{Preface}{Preface}
Your text begins here.
```

The asterisk omits normal pagination and continues lower-case roman numerals. The other commands handle the table of contents and header formatting. Headers are explained in Section 10.4.3.

Following your next \chapter command (this time *without* an asterisk), type

    \pagestyle{headings}

This command returns header formatting to the control of your book or report style.

10.4   Preamble

The *preamble* is a collection of commands issued at the beginning of your input file, which come before the \begin{document} command. In addition to naming the document style and options, the preamble allows you to alter the style's default layout dimensions. Part of the preamble for this book was:

```
% Preamble for LaTeX for Scientists and Engineers
% MHBOOK.STY, MHBK10.STY redesigned by David Buerger
% Last edit September 22, 1989

% document style and options
\documentstyle[ifthen,my,makeidx]{mhbook}
\makeindex

% text dimensions
\topmargin 0mm
\textwidth 27pc
\textheight 44pc

% custom definitions
\newcommand{\mi}[1]{#1\index{#1}}  % indexing command
\newcommand{\bs}{\char '134}  % backslash char. for \tt font
\newcommand{\cir}{\char '136} % circumflex char. for \tt font
\newcommand{\til}{\char '176} % tilde char. for \tt font
\newcommand{\rb}{\char '135}  % right bracket for \rm font
\newcommand{\lc}{\char '173}  % left curly brace for \tt font
\newcommand{\rc}{\char '175}  % right curly brace for \tt font

% BibTeX logo
\def\BibTeX{{\rm B\kern-.05em{\sc i\kern-.025em b}\kern-.08em
    T\kern-.1667em\lower.7ex\hbox{E}\kern-.125emX}}

\begin{document}
```

10.4.1 Document Style Options

Comments following the % signs describe what commands are used in this preamble's content, and include information about the document's contents. The \documentstyle command, discussed in Section 3.2, allows several optional settings which modify the general style. Options are placed within the square brackets as in this example; commas separate multiple options. Commonly used options include 11pt and 12pt, which change the 10-point default type to 11- or 12-point; twocolumn, which creates two columns of text on each page (see Chapter 13); twoside (discussed below), which formats text for printing on both sides of each page; and titlepage, used with the article style to create a separate page for the title with the \maketitle command.

This example's \documenstyle options includes ifthen, a programming-like procedure scheme used with the \renewcommand command to manipulate LaTeX's counter system (see Chapter 14). The my option is a collection of personal macros that I use with most documents. The makeidx option is used to create an index (see Chapter 12).

The document style used for this book is a special version of the standard book style, modified to reflect McGraw-Hill's layout requirements. Since the book is set in 10-point type, the associated BK10.STY file also was changed. LaTeX style file "hacking" can be a complex project, and requires both a knowledge of LaTeX and TeX internal programming, plus help from a knowledgeable document designer.

10.4.2 One-Sided vs. Two-Sided Output

The default page output style for the article and report document styles is one-sided; the default for the book style is two-sided. One-sided output means text is formatted for printing on one side of a page only. Two-sided output means text is formatted to be printed on both sides of a page—odd page numbers appear on the right side and even page numbers appear on the left. If you use the two-sided option, pay close attention to additional page formatting commands—particularly headers (see Section 10.4.3), plus the \oddsidemargin and \evensidemargin settings (see Figure 10.1). These two settings control the amount of space between a spot one inch from the left edge of the page, and the point where the left edge of text appears on odd and even pages. Usually some trial-and-error format-and-print tests must be done to get this setting right.

10.4.3   Headers

The type of header that appears in your document depends upon the document style used. If one-sided printing is used, the book and report styles show the chapter name and page number on each page head; the article style shows the section name and page number. If the two-sided option is used, the book and report options show the chapter name and page number on even page heads, and the section name and page number on odd page heads; the article option shows the section name and page number on even page heads, and the subsection name and page number on odd page heads.

Default headers can be modified with the \pagestyle{ } command. Any one of four options are typed in the curly braces. They include:

plain   Empty header; page number in footer. This is the default page style.

empty   Empty header and empty footer.

headings   Headers are determined by the document style (discussed above).

myheadings Permits custom headers (discussed below).

The preamble command you use to set custom headers depends on what page printing option you use.

- One-sided page headers:

  \pagestyle{myheadings}
  \markright{Custom header here}

- Two-sided page headers:

  \pagestyle{myheadings}
  \markboth{Left-page header here}{Right-page header here}

Headers do not appear on the first page of a document, nor on the first page of new chapters.

10.4.4   Page Layout Options

Page layout dimensions are controlled by settings in LaTeX's document style files. You can change these settings by either modifying the style files, or by issuing precise dimension commands in the input file's preamble. The example in Section 10.4 shows new dimensions for the top margin, text width, and text height. Dimensions can be expressed in many ways, including millimeters, centimeters, inches, points, and picas. A graphical illustration of layout commands for single-column documents is in Figure 10.1. Two-column document preamble commands are illustrated in Table 10.2.[1]

10.5 Appendices

Your appendix or appendices will begin where you type the \appendix command. After this command is given, you should continue to name major sections with the \chapter or \section command. After the input file is processed through LaTeX, however, these subsequent chapters will be titled "Appendix A," "Appendix B," and so on, both at the text location and in the table of contents.

10.6 Bibliography, Glossary, and Index

The creation of a bibliography is discussed in Chapter 11. The glossary and index are discussed in Chapter 12.

10.7 Simplifying Document Assembly by Importing Multiple Input Files

Writers usually prefer to break long documents like books and reports into small files—usually one file for each chapter or section. Small files are easier to manipulate with a word processor or text editor. This technique also makes it faster to debug LaTeX input files and get the whole document looking right.

To show the basic idea, examine the master input file which processed this book:

```
% Main file for LaTeX for Scientists and Engineers
\input{preamble.tex}
\bibliographystyle{plain}
\nocite{chicago}
```

---

[1] Figures courtesy of Nelson H. F. Beebe.

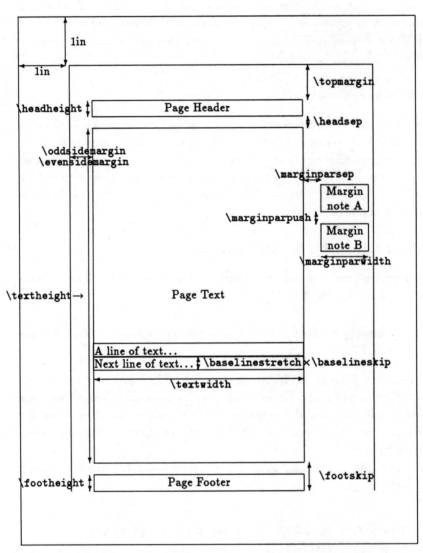

Figure 10.1. LaTeX Single-Column Page Layout Commands.

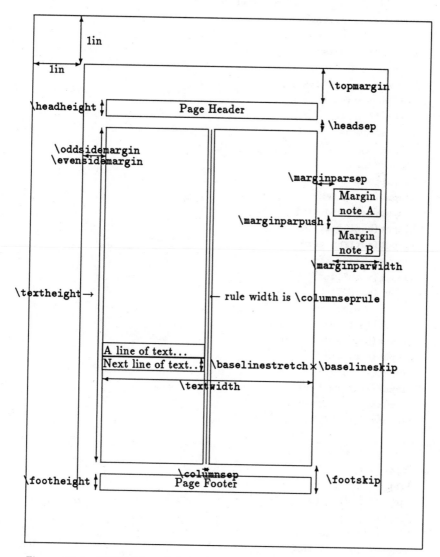

Figure 10.2. LaTeX Two-Column Page Layout Commands.

```
% title page
\input{titlepg.tex}

% copyright page, notices, dedication
\input{notices.tex}

% table of contents
\newpage
\pagenumbering{roman}
\tableofcontents
\listoffigures
\listoftables

% chapters
\input{preface.tex}
\input{intro.tex}
\input{features.tex}
\input{begin.tex}
\input{environ.tex}
\input{change.tex}
\input{math1.tex}
\input{math2.tex}
\input{tables.tex}
\input{foot.tex}
\input{org.tex}
\input{bib.tex}
\input{index.tex}
\input{columns.tex}
\input{special.tex}
\input{errors.tex}

% appendices
\appendix
\input{answers.tex}
\input{samples.tex}
\input{symbols.tex}
\input{fontsize.tex}
\input{glossary.tex}

% bibliography
\bibliography{mybib}
\addcontentsline{toc}{chapter}{Bibliography}
```

```
% index
\cleardoublepage
\printindex
\addcontentsline{toc}{chapter}{Index}

\end{document}
```

Each chapter or informational unit is kept in a separate input file. The first line of each input file has a \input{preamble.tex} command to provide appropriate setup information (see Section 10.4 for an example). The last line of each input file has a \end{document} command. In this manner, each input file functions as its own complete document while text is composed and processed through LaTeX to ensure proper layout.

When you have finished writing and formatting a chapter or section, type a % symbol in front of the \input{preamble} and \end{document} commands. As discussed in Section 3.6, this "comments out" the commands and makes them invisible to LaTeX. Control of chapter input files thus passes to the main input file detailed above. When the main file is processed, each chapter file is automatically consolidated into the book. This procedure also produces all reference numbers, cross-references, proper pagination, glossary, bibliography citations, and index creation.[2]

The \input{ } command adds the contents of the named file to the input file at the point where the command is typed. It does not automatically cause a page break. In the above example, the \chapter commands in each input file automatically generated a page break. In cases where a page break must be manually generated (such as for the index), you can use either the \newpage or \clearpage command. The page break happens at the exact spot where the command is typed. If your document is in \twocolumn mode, \newpage merely starts a new column. In this case, \clearpage will start a new page. If you want to ensure that text starts on a new right-hand page (odd page number), use the \cleardoublepage command. Extra blank pages can be generated if you create "invisible" text on them (e.g. type \mbox{ } before issuing the \newpage command).

The \include{ } command works similarly to the \input{ } command, except that it automatically generates a page break at the point where it is typed.

By using the \input{ } assembly technique, it is easy to assemble small

---

[2] Glossary, bibliography and index creation require extra steps, detailed in Chapters 11 and 12.

files into one large document. Furthermore, should you decide to change parameters that affect the whole document, one change in the common preamble file will flow through to every input file.

Problems

Problem 10.1 How would you type the \documentstyle command to create 12-point, two-column output using the article document style?

Problem 10.2 What command would you type to create custom headers with "Great Research Projects of the 1980s" on left pages, and "Psychosis and Typesetting" on right pages?

Problem 10.3 What commands would you use to customize output such that the text width would be 5 inches, text height would be 7 inches, and the top margin would be 2 inches?

Creating a Bibliography

Most nonfiction writing relies partly upon secondary sources. Citations of secondary articles, books, and manuscript material usually are made in short form within a document. These citations—typically the author and date, plus volume and page numbers where necessary—are fully described in a bibliography at the end of the document.

LaTeX can ease the normal tedium of bibliography construction. Bibliographies can be created in two ways. The first is on a one-by-one basis for each document; the second is a database technique that lets you reuse collected references for any number of projects. While the latter method requires some advance effort to construct a database, it eliminates virtually all work required to create future bibliographies.

The first step to the latter scheme is creating a bibliographic database file containing full reference information to every source cited in the document. This database should also contain keys for each reference you wish to refer to in the body of your text. A key is a phrase typed in the body of your text that ties citations into the bibliographic database. Bibliography and reference citations are automatically created by processing your document through LaTeX, and another program called BibTeX.[1] BibTeX is a publicly available program tailored for use with LaTeX, and is included with most LaTeX distributions.

11.1 Creating a Bibliographic Database

The bibliographic database is stored in an ASCII text file named with a .bib extension (e.g. MYBIB.BIB). Database entries look something like this:

---

[1] This discussion presumes use of BibTeX version 0.99c.

```
@BOOK{van,
     AUTHOR = "van Leunen, Mary-Claire",
     TITLE = "A Handbook for Scholars",
     ADDRESS = "New York",
     PUBLISHER = "Alfred A. Knopf",
     YEAR = "1979"      }
@INPROCEEDINGS{per,
     AUTHOR = "L. M. Pereira",
     TITLE = "''{L}ogic {C}ontrol {W}ith {L}ogic''",
     BOOKTITLE = "Proceedings of the First International
                 Logic Programming Conference",
      YEAR = "1982",
     PAGES = "9--18"    }
@ARTICLE{kow,
     AUTHOR = "J. H. Coombs and A. H. Renear and S. J. DeRose",
     TITLE = "''{Markup Systems and the Future of
                 Scholarly Text Processing}''",
     JOURNAL = "Communications of the ACM",
     YEAR = "1987",
     VOLUME = "30",
     NUMBER = "11",
     PAGES = "933--47" }
```

The word on the first line of each record that follows the entry type @
statement and open curly brace is the "key"; the above example's keys
are van, per, and kow. Keys can contain virtually any character except
commas; they are case-sensitive, so it is wise to stay with one style (e.g. all
lower-case). Fields are delimited with a quote mark characters (") or open
curly brace at the beginning, and an opposite closed character at the end.
Each field also ends with a comma, except for the last field in each record.
Records must be delimited by an open curly brace or parentheses at the
beginning, and an opposite closed character at the end.

Many word processors and text editors can help speed up the database
creation if you create custom macros and templates. A database in *Scribe*
format is virtually identical to BIBTEX's requirements.

Each field will be reproduced according to built-in style conventions. The
booktitle field, for example, will automatically italicize in most styles. If
a publication has multiple authors, you should include the word "and" as
shown above. You can have specific letters or words capitalized in fields
where a style normally formats all letters in lower-case (e.g. title field in
inproceedings entries), by surrounding the text string to be capitalized
with curly braces. The example shows two ways to do this. The same

technique works if you want to include accented characters in bibliographic information. A similar option is illustrated for adding open (' ') and closed (' ') quote marks.

The @BOOK statement means the entry is a book. Many other entry types are allowed; each has required fields that are used in the final bibliography. Optional fields that are used if information is present and ignored if the field is blank, and ignored fields that allow you to record relevant information (e.g. an abstract) that never appears in the printed bibliography. Fields can appear in any order within each record.

11.2 Entry Types

Following are entry types[2] that you can place into your bibliographic database:

article
: Journal or magazine article. Required fields: author, title, journal, year. Optional fields: volume, number, pages, month, note.

book
: Book with an explicit publisher. Required fields: author or editor, title, publisher, year. Optional fields: volume or number, series, address, edition, month, note.

booklet
: Printed and bound publication without a named publisher or sponsoring institution. Required field: title. Optional fields: author, howpublished, address, month, year, note.

conference
: Same as inproceedings; included for *Scribe* compatibility.

inbook
: Chapter and/or a range of pages in a book. Required fields: author or editor, title, chapter and/or pages, publisher, year. Optional fields: volume or number, series, type, address, edition, month, note.

incollection
: Part of a book with its own title. Required fields: author, title, booktitle, publisher, year. Optional

---

[2] Adapted courtesy of Oren Patashnik, Stanford University, who implemented the BibTeX program and wrote its documentation. His complete documentation, called "BibTeXing," is included with the publicly available BibTeX distribution.

fields: editor, volume or number, series, type, chapter, pages, address, month, note.

inproceedings Conference proceedings article. Required fields: author, title, booktitle, year. Optional fields: editor, volume or number, series, pages, organization, publisher, address (used to cite conference location), month, note.

manual Technical documentation. Required field: title. Optional fields: author, organization, address, edition, month, year, note.

mastersthesis Master's thesis. Required fields: author, title, school, year. Optional fields: type, address, month, note.

misc Used when nothing else fits. Required fields: none. Optional fields: author, title, howpublished, month, year, note.

phdthesis Ph.D. thesis. Required fields: author, title, school, year. Optional fields: type, address, month, note.

proceedings Conference proceedings. Required fields: title, year. Optional fields: editor, volume or number, series, address (used to cite conference location), month, organization, publisher, note.

techreport Report published by a school or other institution, usually numbered within a series. Required fields: author, title, institution, year. Optional fields: type, number, address, month, note.

unpublished Document with an author and title, but not formally published. Required fields: author, title, note. Optional fields: month, year.

11.3   Fields

Following is a list of fields required by many bibliography styles.[3] Standard bibliography styles are described in Section 11.6. Some fields are required by certain entry types, others are optional. A few fields are ignored by

---

[3] Adapted courtesy of Oren Patashnik.

entry types; their purpose largely is for informational notes not printed in the final document.

address	Publisher's address. List just the city for major publishers, the complete address for small publishers.
annote	Listing used only for nonstandard annotated bibliography styles.
author	Name(s) of the author(s).
booktitle	Book title, part of which is being cited. Use `title` field for book entries.
chapter	Chapter number.
edition	Book edition (e.g. "second").
editor	Editor name(s). If item includes an **author** field, the **editor** field lists the editor of the book or collection in which the reference appears.
howpublished	How nonstandard item was published.
institution	Institution that published the work.
journal	Journal name. Computer-science abbreviations are provided for many journals (listed below).
key	Used to alphabetize and create a label when the **author** and **editor** fields are missing. This field should not be confused with the key that appears in the `\cite` command and at the beginning of the entry.
month	Month when work was published or, for an unpublished work, when it was written.
note	Additional information that can help the reader.
number	Number of journal, magazine, or technical report. An issue of a journal or magazine is usually identified by its volume and number; organizations that issue technical reports usually assign a number.
organization	Organization sponsoring a conference.

pages
: Page number or range of numbers such as 12–56, 4,23,33–45, or 78+ (second example shows following pages that aren't in a simple range). BibTeX automatically converts a single dash into a double en-dash (--) appropriate for numeric ranges.

publisher
: Publisher's name.

school
: Name of the school where a thesis was written.

series
: Name of a series or set of books. When citing an entire book, the **title** field gives its title and an optional **series** field gives the name of a series in which the book is published.

title
: Work's title, typed as explained above.

type
: Type of a technical report (e.g. "Research Note").

volume
: Volume of a journal or multivolume work.

year
: Year of publication or, for an unpublished work, the year it was written. This field's last four characters should contain only numerals (a listing such as "about 1945" is permissible).

11.4   Automatic Journal Citations

The standard bibliography style files define automatic use of the following computer science-related abbreviations, used in BibTeX's appropriate fields:

acmcs
: *ACM Computing Surveys*

acta
: *Acta Informatica*

cacm
: *Communications of the ACM*

ibmjrd
: *IBM Journal of Research and Development*

ibmsj
: *IBM Systems Journal*

ieeese
: *IEEE Transactions on Software Engineering*

ieeetc
: *IEEE Transactions on Computers*

`ieeetcad`   *IEEE Transactions on Computer Aided Design of Integrated Circuits*

`ipl`   *Information Processing Letters*

`jacm`   *Journal of the ACM*

`jcss`   *Journal of Computer and System Sciences*

`scp`   *Science of Computer Programming*

`sicomp`   *SIAM Journal on Computing*

`tocs`   *ACM Transactions on Computer Systems*

`tods`   *ACM Transactions on Database Systems*

`tog`   *ACM Transactions on Graphics*

`toms`   *ACM Transactions on Mathematical Software*

`toois`   *ACM Transactions on Office Information Systems*

`toplas`   *ACM Transactions on Programming Languages and Systems*

`tcs`   *Theoretical Computer Science*

11.5   How to Cite References in the Text

Citing bibliographic references is similar to the way you cite cross-references described in Section 9.2. For example, assume your bibliography contains three references with the following key fields: `lions`, `tigers`, and `bears`. This is how you would cite these references:

Dorothy would have better prepared for her adventures in the Land of Oz had she first consulted [2,3], and especially [1, pp. 46–79].

```
Dorothy would have better
prepared for her adventures
in the Land of Oz had she
first consulted \cite{lions,
tigers}, and especially
\cite[pp.\ 46--79]{bears}.
```

Note that multiple references are cited by separating each key with a comma. Ranges of page numbers are cited in optional brackets preceding the reference key. In this example, reference numbers that appear in the final text are generated by BibTeX according to a user-specified bibliography style file described in Section 11.6.

11.6   Using BibTeX to Create a Bibliography

Before generating a bibliography you must insert two commands in your document. First, go to the position in your document where you wish to generate the bibliography and type \bibliography{mybib}. In this example, mybib is the name of your bibliographic database file that has the .bib extension. The second step is to select the bibliography style you wish to have your bibliography formatted with by typing this command in your document after the \begin{document} command:

```
\begin{document}
\bibliographystyle{plain}
```

The plain bibliography style is one of four such styles distributed with the current version (0.99c) of BibTeX; each style file ends with a .bst extension. The styles include:

plain   Follows conventions outlined by Mary-Claire van Leunen in *A Handbook for Scholars* [13]. The bibliography is sorted alphabetically by author, then year, then title; each entry is labeled sequentially with numbers. These numbers appear where citations are made in the finished text.

unsrt   The unsorted style is identical to the plain style except that entries are sorted in the order of their first citation, not alphabetically by author.

alpha   Identical to the plain style except instead of numbers, a combination of the author's name and year of publication is used (e.g. "Knu86"). Sorting order is by label, then author, then year, then title.

abbrv   Identical to the plain style except first names, months, and journal titles are abbreviated.

Other styles included with some distributions are ieeetr, acm, siam, and apalike. See your *Local Guide* for more information on optional styles that might be supported by your local LaTeX administrator.

Once these two commands are issued, you can generate the bibliography. This usually is done about the time you are finishing writing the document. The steps are:

1. Process your document through LaTeX; this generates the .aux file that BibTeX needs to create a .bbl file, and places '??' at each \cite location.

2. Process your document through BIBTEX. On many systems, you run your document through BIBTEX by typing `BIBTEX filename`, (without any extension) followed by the return key. Some systems require you to type `BIBTEX`, then prompt you for the filename. This will create a new `.bbl` file.

3. Process your document through LATEX again; this incorporates the bibliography in your document.

4. Finally, process your document through LATEX a third time; this will handle all forward references in the document.

The resulting `.dvi` file can then be previewed or sent to an output device. In some cases, you might want to include items in your bibliography that are not explicitly cited in the text. In this case, you must use the `\nocite{}` command which generally follows the `\bibliographystyle` command. If you had bibliographic citation key fields called adams and sagan, but no in-text citations, you would include them in your bibliography as follows:

```
\begin{document}
\bibliographystyle{plain}
\nocite{adams}
\nocite{sagan}
```

If you want to create a general reference by including all citations from your bibliographic database, but have no in-text citations, use the `\nocite{*}` command like this:

```
\begin{document}
\bibliographystyle{plain}
\nocite{*}
```

When you use the unsrt style, which lists references in order of their first citation, place the `\nocite{*}` command at the beginning of your input file (as in the above illustration).

11.7 Making a Bibliography Without BibTeX

If you have just a few references, you might want to create a bibliography without using the BIBTEX program. This is done with the thebibliography environment. This environment is similar to the enumerate and itemize list-making environments, with the following exceptions:

- Instead of beginning each bibliography entry with an \item command, you use a \bibitem command.

- The thebibliography environment contains a label width argument that specifies the printing width (in characters) of the widest item label found in the source list.

In-text citations are made the same as above, using the \cite{ } command. The bibliography information is not typed in a database file; instead it is typed within the thebibliography environment, starting at the position in the input file where you want the bibliography to begin. Following is some sample input:

References

[1] Donald E. Knuth. *The TEXbook*. Reading, Ma.: Addison-Wesley Publishing Co.

[2] Sina Spiker. *Indexing Your Book: A Practical Guide for Authors.* Madison, Wisc.: Univ. of Wisc. Press, 1954.

[3] Mary-Claire van Leunen. *A Handbook for Scholars.* New York: Alfred A. Knopf, 1979.

```
\begin{thebibliography}{99}

\bibitem{knuth} Donald E. Knuth.
{\it The \TeX{}book}.  Reading,
Ma.: Addison-Wesley Publishing
Co.

\bibitem{spiker} Sina Spiker.
{\it Indexing Your Book: A
Practical Guide for Authors}.
Madison, Wisc.: Univ. of Wisc.
Press, 1954.

\bibitem{van} Mary-Claire van
Leunen. {\it A Handbook for
Scholars}.  New York: Alfred A.
Knopf, 1979.

\end{thebibliography}
```

All \cite{ } references must use keys listed in the bibliography. The "References" title is automatically produced by the thebibliography environment. You can override the default numbering system with optional labels used as follows:

[Knuth 86] Donald E. Knuth. *The TEXbook* . . .

```
\bibitem[Knuth 86]{knuth}
Donald E. Knuth. {\it The
\TeX{}book}...
```

Your input file must be processed through LaTeX twice to produce the final bibliography. You may want to use the \newpage command to separate the bibliography from the last chapter or section, and start it on a new page.

Problems

Problem 11.1 Create bibliography database entries for the following three titles:

- *America—20th Century Poetry: Landscapes of the Mind.* McDougal, Littell & Co., Evanston, Ill., 1973.

- Edward Connery Lathem, editor. *The Poetry of Robert Frost.* Holt, Rinehart and Winston, New York, 1969.

- Sina Spiker. *Indexing Your Book: A Practical Guide for Authors.* Univ. of Wisc. Press, Madison, Wisc., 1954.

Problem 11.2 How would you cite references to the three items created in Problem 11.1?

Problem 11.3 How would you create a bibliography database entry for the following item?

- William Strunk, Jr. and E. B. White. *The Elements of Style.* The Macmillan Co., New York, 1972.

Creating a Glossary and Index

12.1 Glossary

LaTeX has no environment expressly designed to make a glossary. The easiest method to make a one-column glossary is by starting it with a \chapter command (for a book) or a \section command (for an article). Entries can be formatted with the description environment (see this book's glossary for an example). If you want to omit the chapter or section number from the glossary's title page, use this command:

```
\chapter*{Glossary}
\addcontentsline{toc}{chapter}{Glossary}
\pagestyle{myheadings}
\markboth{Glossary}{Glossary}
```

The \addcontentsline command ensures that the glossary is listed in the table of contents. The \pagestyle and \markboth commands will ensure that "Glossary" appears on each page's header.

12.2 Index

Traditionally, the creation of an index has been a laborious, last-minute effort done with 3 × 5 index cards and lots of patience. LaTeX provides the means to automate the creation of an index. If you make an index with LaTeX alone, however, the result is just a text file that requires substantial editing before it is presentable. When used with a publicly available supplementary indexing program called *MakeIndex*, however, it is easy to produce a first-class index. These instructions assume that you will use *MakeIndex*.

The purpose of an index is to help readers quickly find information. In order to get a reader's perspective, it generally is wise to start indexing a document after it is completed. A useful technique is to use a sorting program that produces an alphabetical list of all words in your document. This list will help you notice which words are important to index. It also will help you keep all index entries and cross-references uniform within the index. See [10] for other suggestions on index creation.

The creation of an index requires two steps: (1) marking each word or words you want to include in the index, and (2) running the *MakeIndex* program to create the index.

12.2.1   Marking Words to Be Indexed

This section shows how you mark words in the input file for indexing. The basic LaTeX index command is used like this:

```
The Rolls Royce\index{Rolls Royce} is made in England.
```

As you can see, this involves extra typing. The \newcommand command lets us create a shortcut macro (see Section 14.1 for information). In your document preamble, type the following command:

```
\newcommand{\mi}[1]{#1\index{#1}}
```

The new command, \mi (short for "my index"), now allows you to mark index entries like this:

```
The \mi{Rolls Royce} is made in England.
```

All of the following illustrations that use \mi assume that you use this \newcommand definition.

Continuing this automobile example, if your document has four references to "Rolls Royce" marked for indexing, the resulting index would look like this:

Rolls Royce, 3, 14, 76, 105

Two identical items marked on the same page would receive only one page reference in the index.

You also can produce sub-entries, sub-sub-entries, and cross-references. Sub- and sub-sub-entries are created by separating them from the main entry with the exclamation ! symbol. Cross-references are separated with the vertical bar | symbol. For example,

      Maserati, 50
      Rolls Royce, 3, 14, 76, 105
          manufacturing techniques of, 77
          popularity of, 5
      sports cars, 13, 44
          racing, *see* Maserati
              corporate participation in, 51

was created with this input:

```
p. 5:  Rolls Royce\index{Rolls Royce!popularity of}
p. 13: \mi{sports cars}
p. 44: \mi{sports cars}
p. 50: \mi{Maserati}
p. 50: sports cars\index{sports cars!racing|see{Maserati}}
p. 51: sports cars\index{sports cars!racing!corporate participation in}
p. 77: Rolls Royce\index{Rolls Royce!manufacturing techniques of}
```

Note that whenever a sub-entry, sub-sub-entry, or cross-reference index command is typed, you must use the normal \index command instead of the \mi redefinition. Otherwise the entire command will be reproduced in the final document. The cross-reference "see" command should only be used once for each entry to prevent double occurrences in the index.

MakeIndex also allows you to specify a page range for an index entry. Ranges are specified by placing a \index{...|(} command at the range beginning and a \index{...|)} command at the range ending. This sample output,

      Detroit, 86
      Ford Motor Co., 86
         Mustang, 86–92
         Pinto, 92–93
      General Motors Corp., 101–109

was produced with this input:

```
p. 86:  \mi{Detroit}
p. 86:  \mi{Ford Motor Co.}
p. 86:  Ford Motor Co.\index{Ford Motor Co.!Mustang|(}
p. 92:  Ford Motor Co.\index{Ford Motor Co.!Mustang|)}
p. 92:  Ford Motor Co.\index{Ford Motor Co.!Pinto|(}
p. 93:  Ford Motor Co.\index{Ford Motor Co.!Pinto|)}
p. 101: General Motors Corp.\index{General Motors Corp.|(}
p. 109: General Motors Corp.\index{General Motors Corp.|)}
```

Make sure that you use the proper delimiter—e.g. (vs.)—in marking the start and end of range limits.

A common error made using *MakeIndex* is adding extra spaces within an argument. Assuming the ␣ character represents a space, the following index markers would produce separate entries in the final index:

- \mi{Mustang}

- \mi{␣Mustang}

- \mi{Mustang␣}

- \mi{␣␣Mustang}

All index entries are case sensitive. Index commands for Mustang and mustang would produce two separate entries in the final index.

It is best to place index entries underneath section headings, or at the beginning or end of paragraphs. Otherwise you may notice extra space created between words. If you must place an index entry within a paragraph, do not create more than one per line. Never stack index entries on two consecutive lines within paragraphs; otherwise you will create extra space between words. A single index entry must be on the same line of input.

You may want your index to reference a concept that cannot be explicitly linked to a word contained in your text. This is done by typing the \index command at the place the concept occurs in your input file. Use the normal index command sequence of \index{...!...!...|see{ }} instead of the new command \mi. This technique can index ideas on single pages as well as in page ranges.

The final index's entries are arranged in alphabetical order, with entries separated by letter groups. *MakeIndex* will place all entries that have special typefaces or other formatting commands (e.g. the LaTeX symbol), as well as all symbols marked for indexing, at the beginning of the index. See this book's index for examples. *MakeIndex* will not index entries containing the \verb command.

12.2.2   Using MakeIndex to Create an Index

Before running *MakeIndex*, you must take care of three things:

- Add the makeidx option to your \documentstyle command. For example, if you are using the 11-point book style, the command would look like this:

```
\documentstyle[11pt,makeidx]{book}
```

- Just after the \documentstyle command, add this command:

```
\makeindex
```

- Just before the \end{document} command, add these commands:

```
\cleardoublepage
\printindex
```

The first command causes the index to start on an odd page number. The second command causes the index to be printed at this spot.

You must process your input file once through LaTeX to create an .idx file. Assuming the resulting file is called MYFILE.IDX, type this command:

```
makeidx myfile.idx  <return>
```

This produces a file called MYFILE.IND. By processing MYFILE.TEX through LaTeX one more time, *MakeIndex* will produce a finished index at the end of your document.

MakeIndex is a program written in "C" by Pehong Chen, with consultation provided by Michael Harrison and Leslie Lamport. It was modified for portability by Nelson H. F. Beebe; free source code is available on the standard TeX distribution tapes from Stanford University. Source files also are available via FTP Internet access in Stanford's <TEX.*> directories at score.stanford.edu.

MakeIndex source code is also available for a nominal fee from VorTeX Distribution, EECS—Computer Science Division, 571 Evans Hall, University of California at Berkeley, Berkeley, California, 94720; telephone (415) 642-1469. E-mail inquiries may be sent to vortex@ucbarpa.berkeley.edu.

More in-depth information on *MakeIndex* is in the documentation file written by Leslie Lamport, included with the program distribution.

Problems

Problem 12.1 How would you mark the italicized words in this paragraph for indexing with the MakeIndex program, cited from [11, p. 17]?

> Vigorous writing is concise. A *sentence* should contain no unnecessary words, a *paragraph* no unnecessary sentences, for the same reason that a drawing should have no unnecessary lines and a machine no unnecessary parts. This requires not that the *writer* make all his sentences short, or that he avoid all detail and treat his subjects only in outline, but that every word tell.

Problem 12.2 How would you mark the phrase "Vigorous writing" for indexing such that the index would read "writing, vigorous"?

Problem 12.3 How would you mark the phrase "Vigorous writing" for indexing such that the index would read "writing, vigorous, *see* Chapter 12"?

Problem 12.4 How would you mark a range of pages discussing the topic "vigorous writing" for reference in the index?

Two-Column Documents

13.1 The "Twocolumn" Option

You can format your entire document in two-column mode by invoking the twocolumn option in your preamble's document style declaration as follows:

```
\documentstyle[twocolumn]{article}
```

Be sure to omit any extra spaces inside the square brackets.

13.2 The "Twocolumn" Proceedings Option

Most conference proceedings editors require that you submit articles pasted up in two columns on 11 × 14 inch mats. Editors then photo-reduce these sheets and duplicate them for distribution. If your proceedings editor insists that you use this format, you should create your LaTeX document in single-column format, with a text width conforming to your proceedings' requirement (usually $4\frac{1}{8}$ inches). Use the 12-point document style option to offset the editor's reduction of your originals.

A number of editors also accept laser-printed submissions created on $8\frac{1}{2}$ × 11 inch paper. Most LaTeX distributions include a proc document style option that creates two-column output tailored for ACM and IEEE conference proceedings.

The proc document style option is designed to work with the article document style. You invoke it by typing the following in your preamble's document style command:

```
\documentstyle[proc]{article}
```

Other options can be used with "proc" if desired.

When using the proc option, you need to place a \copyrightspace command just after your \begin{document} command. This will create a blank footnote at the bottom of the first column on the first page. Your proceedings editor will place the copyright notice in this position. If you have a \footnote command that places a footnote in the first column, the \copyrightspace command should come after the footnote.

If you wish to place your name and title header on each page, type this command in your preamble before the \begin{document} command:

```
\markright{Name---Header Title}
```

Substitute your name for "Name" and your paper's abbreviated title for "Header Title." This will be printed at the bottom of each page.

13.3   Switching to Two-Column Mode in Text Body

LaTeX allows you to switch to two-column mode in the middle of your text's body by typing the \twocolumn declaration. This automatically executes a \clearpage command that causes subsequent text to be formatted in two columns on a new page. A \onecolumn declaration again executes the \clearpage command and switches text back to one-column format on a new page.

If you execute a \newpage command while in either the twocolumn style or after invoking the \twocolumn command, text will start in a new column rather than on a new page.

The \clearpage command is built into LaTeX's \twocolumn command. It is possible to change a style file to allow switching to and from two-column mode in the middle of text body. Such style file modifications, however, require advanced knowledge of LaTeX and TeX, and are beyond the scope of this book.

13.4   Parallel Text With the "Minipage" Environment

Another way to place text side-by-side is to use the minipage environment. This environment creates a special type of LaTeX *parbox*—lines of text controlled outside the normal left-right (LR) text mode. (Two examples of a parbox are the figure and table environments.) The minipage environment lets you create two parallel "pages," the size of which are controlled separately from the actual page on which the parallel parboxes appear.

An example of a minipage environment is seen in the following list of *dBASE III Plus* commands:

```
SET DEFAULT TO b          Set drive B as the default drive
USE testfile              Retrieve file named TESTFILE.DBF
DISPLAY STRUCTURE         List all fields, their types and widths
```

This example was created with six concatenated `minipage` environments as follows:

```
\begin{flushleft}
\begin{minipage}[t]{45mm}
{\tt SET DEFAULT TO b}
\end{minipage}
\begin{minipage}[t]{65mm}
Set drive B as the default drive
\end{minipage} \\
\begin{minipage}[t]{45mm}
{\tt USE testfile}
\end{minipage}
\begin{minipage}[t]{65mm}
Retrieve the file named {\tt TESTFILE.DBF}
\end{minipage} \\
\begin{minipage}[t]{45mm}
{\tt DISPLAY STRUCTURE}
\end{minipage}
\begin{minipage}[t]{65mm}
List all fields, their types and widths
\end{minipage}
\end{flushleft}
```

In this example, the `minipage`'s were set up in pairs. Distance specifications (in curly braces) can be varied to adjust the width of each parallel parbox. You need to coordinate the width of each parbox to ensure that they don't take up more room than your full-page column width allows. A minipage can include multiple lines. It even can have its own footnotes, which are attached directly to the minipage and are not included with normal footnotes. The \\ commands seen in this example push following pairs of minipages to the next line; this lets you stack them as desired. The optional [t] stands for "top," and ensures that matching minipage parboxes are aligned at the top.

13.5   Creating Newsletters With LaTeX

Creating newsletters is not LaTeX's forte. Nevertheless, it is possible to create two-column newsletters with LaTeX. The simplest method I've found to create a banner across the top of the first page, followed by two-column text underneath, is to use the following sample preamble outline:

```
\documentstyle[twocolumn]{article}
\topmargin 0pt
\headsep 0pt
\headheight 0pt
\textheight 8.8in
\textwidth 6.5in
\columnsep 20pt
\columnseprule .5pt
\setlength{\parskip}{.5em}
\begin{document}
\title{\Large\begin{center}{\it UNIVERSITY OF ILLINOIS} \\
\vspace{2 mm}
\Huge
{\bf COMPUTER SCIENCE DEPARTMENT NEWS} \\
\end{center}
\vspace{6 mm}
\small
The monthly student newsletter published by
    \hfill Noreen Edwards, Editor \\
the Computer Science Dept. \hfill
    November 1990
\vspace{1 mm}
\hrule}
\author{}
\date{}
\maketitle
```

The newsletter's body follows this banner in two-column mode. These parameters create a two-column newsletter on $8\frac{1}{2} \times 11$ inch paper. The banner is created by manipulating the title page commands discussed in Section 10.2. You can experiment with the example to create different banners.

Problems

Problem 13.1 What command do you type, and where is it placed, if you want to format a document entirely in a two-column format?

Problem 13.2 What command do you use to format a document in the two-column ACM and IEEE conference proceedings style?

Problem 13.3 What commands do you use to switch from one- to two-column format in mid-text, and then switch back to one-column mode?

Problem 13.4 Create the following text with the minipage environment:

> Beginning LaTeXers would do well to take counsel from the "Grand Wizard of TeX," Donald Knuth:[a]
>
> When you first try to use TeX, you'll find that some parts of it are very easy, while other things will take some getting used to. A day or so later, after you have successfully typeset a few pages, you'll be a different person; the concepts that used to bother you will now seem natural, and you'll be able to picture the final result in your mind before it comes out of the machine.

> [a]Cited from *The TeX book*, p. vi.

Chapter 14

Special Operations

Typesetting technical documents can be a complex job. Like most typesetting systems, LaTeX has its own collection of "utility commands" that allow you to tweak your document's appearance. This chapter discusses some of these special operations and how you can use them to make LaTeXing easier.

14.1 Creating Macros With "Newcommand" and "Renewcommand"

The \newcommand command is primarily used for creating shortcuts to typing repetitive text or commands. For example, suppose you had to write a paper that required frequent full-name references to the *Journal of Management Information Systems*. You could either type this name repeatedly, or use \newcommand like this:

```
\newcommand{\jmis}{\it Journal of Management
                        Information Systems\/}
```

The journal's full name will appear any place you type \jmis. The command's format is simple: the argument contained in the first pair of curly braces is the macro code; the second argument is what actually appears when you type the first argument.

The \newcommand command can also be used to redefine frequently used commands into short abbreviations. For example, \begin{itemize} and \end{itemize} could be shortened respectively to \bi and \ei like this:

```
\newcommand{\bi}{\begin{itemize}}
\newcommand{\ei}{\end{itemize}}
```

Be careful using \newcommand if you are combining redefinitions with the math environment. If you have a need to frequently use Greek symbols in roman text, you can redefine the math mode Greek letters like this:

```
\newcommand{\al}{$\alpha$}
\newcommand{\be}{$\beta$}
\newcommand{\ga}{$\gamma$}
```

In this example, use of \al, \be, and \ga in roman text would respectively produce α, β, and γ. If you use these macros *within* the math mode, however, the embedded $ will switch text back to roman. You would thus need two sets of macros to produce the same characters in both roman and math mode.

The \newcommand command also has optional replacement arguments that are useful with varying elements. Suppose you need to type repetitive math expressions in roman text, as in this example:

If $a_0 = 38$ and $a_1 = 19$, we can obtain values for the following table.

This particular expression is automated with this \newcommand:

```
\newcommand{\im}[3]{{$#1$}{$_#2$}{$=#3$}}
```

The \newcommand command sequence begins with the macro name, then the number of replacement variables (up to a maximum of nine), then instructions for how each variable is to be formatted. Each variable is coded with a #n where n is the variable number. Once you've typed the \newcommand, the input file would look like this:

```
If \im{a}{0}{38} and \im{a}{1}{19}, we can obtain
values for the following table.
```

Another related command is \renewcommand{ }{ }, used to modify behavior of previously defined commands. The first parameter is the command to be changed; the second parameter is what you are changing the command into. For example, if you used boldfacing for emphasis in a document, and decided to try italicizing for emphasis instead, you would place this command in your document's preamble:

```
\renewcommand{\bf}{\it}
```

Every instance of the \bf command would be treated as if it were \it instead.

Should you use macro redefinitions, place them in your input file's preamble. This placement makes it easier to modify them if changes are required. Also note that while macros can make typing text faster, they could lead to problems if you trade text with other people in co-authoring situations. If you trade text, you must also trade macros; otherwise the input file won't properly format on other people's computers. Be sure that whatever macros you do use are well-documented. This will ensure that you don't forget what the macros do, and will help other people understand what you are doing if they must use your text.

14.2 Creating Custom Environments With "Newenvironment" and "Newtheorem"

The \newenvironment command lets you define custom environments. For example, if you frequently cited poetry and wished to set every poem in emphasized type, \newenvironment could be used like this:

```
\newenvironment{mypoem}{\begin{verse} \em}{\end{verse}}

The following Robert Frost poem (\cite[p.\ 247]{frost})
was first published in 1930:

\begin{center}Devotion\end{center}
\begin{mypoem}
The heart can think of no devotion\\
Greater than being shore to the ocean---\\
Holding the curve of one position\\
Counting an endless repetition.
\end{mypoem}
```

Note how the \em command is placed within the \begin argument. An \end argument always terminates a special condition started within an environment. The end result is as follows:

The following Robert Frost poem ([7, p. 247]) was first published in 1930:

Devotion

> *The heart can think of no devotion*
> *Greater than being shore to the ocean—*
> *Holding the curve of one position*
> *Counting an endless repetition.*

The \newenvironment command also lets you use replacement variables, just like \newcommand.

Mathematicians, logicians, and philosophers (among others) frequently need to discuss ideas within the context of logical structures, such as axioms, propositions, and theorems. The \newtheorem environment lets you define custom environments best-suited to these needs. Each environment is numbered consecutively from the beginning of your document. For example, if you want to define a environment called myth (for "mytheorem") that labels ideas as "Theorem," type the following command in your document preamble:

```
\newtheorem{myth}{Theorem}
```

Here is an example of output produced with this "myth" environment:

Theorem 1 *This is the first theorem. Its title is automatically numbered and formatted in boldface. The theorem's body automatically prints in italic typeface.*

Theorem 2 *This is the second theorem. Everything prints the same as before, but the theorem's number automatically increases by one.*

This example's input was as follows:

```
\begin{myth}
This is the first theorem.  Its title is automatically
numbered and formatted in boldface.  The theorem's
body automatically prints in italic typeface.
\end{myth}

\begin{myth}
This is the second theorem. Everything prints the same
as before, but the theorem's number automatically
increases by one.
\end{myth}
```

The \newtheorem environment has an additional numbering argument (in square braces) that lets its numbering scheme be patterned after a specific sectional unit like chapter or section. The following command sets up a custom environment called myprop (for "my proposition"), with numbering based on chapter numbers.

```
\newtheorem{myprop}{Proposition}[chapter]
```

If you were working on Chapter 7, for example, and typed the following:

    \newtheorem{myprop}{Proposition}[chapter]

    \begin{myprop}
    This is the first proposition in Chapter 7. Apart from
    numbering consecutively within the chapter, its format
    is identical to the prior example.
    \end{myprop}

This input would produce:

Proposition 7.1 *This is the first proposition in Chapter 7. Apart from numbering consecutively within the chapter, its format is identical to the prior example.*

14.3 Numbering Counters

Occasionally you may wish to alter the value or appearance of LaTeX's numbering. Numbering schemes are based on values kept by LaTeX in various *counters*. Every automatically generated value is based on a counter. Values of each counter can be changed by using several commands described below. These commands require that you specify the counter name for the value you wish to alter. Table 14.1 contains a list of each LaTeX sectional unit and environment with its corresponding counter name.

Counter values may be altered with the \setcounter command. This command resets the current value of any counter listed in Table 14.1 to a specified new value. Subsequent incremental changes to this counter are based on the new value as a starting point, unless you change that counter to yet another value. For example, to change the value of a new chapter to 8, you would type this command just before the new \chapter command:

    \setcounter{chapter}{7}

The new \chapter command will then increase the value of the **chapter** counter by one to the desired value of eight. Subsequent chapters will increase from the "8" value unless you again change the counter.

Counter changes can also be made with the \addtocounter command. For example, if an equation counter is 12, and you want the next equation number to be 17, you would type the following command just before the next equation:

    \addtocounter{equation}{4}

This will bring the **equation** counter to 16; the next equation increases this new base to the desired value of 17. Negative numbers can be used with this command to reduce counter values.

Table 14.1. LaTeX Sectional Units and Environments With Corresponding Counters

Unit Name	Counter Name
chapter	chapter
enumerate 1st level	enumi
enumerate 2nd level	enumii
enumerate 3rd level	enumiii
enumerate 4th level	enumiv
equation	equation
figure	figure
footnote	footnote
minipage footnote	mpfootnote
page	page
paragraph	paragraph
part	part
section	section
subparagraph	subparagraph
subsection	subsection
subsubsection	subsection
table	table

The **page** counter is slightly different from other counters in that the \setcounter sets the value of the *current* page. Thus, if you type this command anywhere on a page:

    \setcounter{page}{74}

the page will be paginated with "74" when it is processed through LaTeX. Subsequent pages will increase from 74.

The appearance of numbers can also be changed. For example, a **page** counter value "7" can can be displayed as follows:

Command	*Result*
\arabic{page}	7
\Roman{page}	VII
\roman{page}	vii
\Alph{page}	G
\alph{page}	g

Changing the appearance of counter values is done by adding \the in front of a sectional unit you wish to redefine, and by using \renewcommand (and in some cases \newcommand). For example, one way you could change

LaTeX's default numbering scheme to sequentially number pages within chapters (e.g. 1–1, 1–2, 1–3, 2–1, 2–2, 2–3, etc.) is with the following commands:

```
\renewcommand{\thepage}{\arabic{chapter}--\arabic{page}}
\newcommand{\mychapter}[1]{\chapter{#1}\setcounter{page}{1}}
```

The first command, which should be typed in the preamble, redefines the pagination to reflect the new numbering scheme. The \mychapter command resets chapter pagination to a value of one, so it would need to be issued at the beginning of each chapter. Of course, you could also type the more generic command, \setcounter{page}{1} at the beginning of each chapter.

Finally, you may wish to create non-standard counters for special operations. To do this, type the following command:

```
\newcounter{yourcountername}
```

Like the counters discussed earlier, non-standard counters can be increased or otherwise manipulated.

As you experiment with these procedures, you will find different ways to achieve the same effect, the use of which usually becomes a matter of personal preference.

14.4 Combining Graphics and Text

LaTeX does not provide strong support in merging graphics and text. If you need this capability you have four choices: use a different software package; use LaTeX's picture environment; reserve space in LaTeX documents to manually paste in graphics generated with specialized graphics software; or electronically merge the latter graphics with LaTeX documents. Let's discuss the pros and cons of each option in order.

There is little question that graphics packages available on personal computers like the Apple Macintosh make it very easy to generate sophisticated graphs, charts, and other pictorial representations of ideas. Should the lack of strong graphics support in LaTeX make you consider some other means of electronically producing documents? If your primary goal is to produce complex technical documents, especially those which contain lots of math expressions, equations, and other intricate number displays, the discussion in Chapter 1 on the pros and cons of visual vs. markup document production systems should be enough to convince you to work within the confines of LaTeX. If you feel you must use something else, be prepared to spend lots of time visually formatting documents and becoming an expert at equation layout.

LaTeX does provide a native tool for some graphic expression called the picture environment. This environment is not easy to use, and requires lots of manual layout and trial-and-error efforts to make it work. It is based on positioning lines, circles, boxes, and text in specified x and y coordinates, and thus requires some dexterity with coordinate geometry. Examples of what you can produce with the picture environment are Tables 10.1 and 10.2. Since the picture environment is native to LaTeX, its images can be printed with any .dvi translator. The downside, of course, is that such images are very difficult and time-consuming to produce. Images that you would rather use (such as graphs produced with Lotus 1-2-3) cannot be translated into picture environment mode. Given these variables, the picture environment is not recommended for serious graphic development. If you would like to experiment with this tool, you should spend some time with [6, pp. 101–11, 196–99] and make your own judgment.

The third method mentioned—reserving space in LaTeX documents and manually pasting in graphics produced with other software—is the easiest and least painful way to satisfy your need. This technique is merely a variation of the figure environment discussed in Section 8.3. This lets you reserve a specified amount of space that will be floated to its optimal position in your document. For example, if you wanted to paste in a graph that is five inches high, you would type the following:

```
\begin{figure}
    \vspace{5in}
    \caption{This is the figure's caption}
\end{figure}
```

In this example, the optional caption is placed beneath the figure. If you type the \caption command before the \vspace command, the caption will appear above the figure. Standard figure placement options listed in Table 8.2 can be used if desired. After you print the LaTeX document, you simply paste the graphic onto the reserved space and photocopy the page; the photocopy becomes the "original."

The electronic method[1] of merging graphics with LaTeX documents requires a PostScript-compatible laser printer or phototypesetter. The graphic image must be stored in PostScript-encapsulated form. Images can be

---

[1] I am indebted to Nelson H. F. Beebe for help with the following discussion, which is adapted with permission from his *Local Guide, Using LaTeX at the University of Utah.*

merged with TeX's native PostScript \special command. This discussion presumes some familiarity with PostScript programming.

The \special command must be in one of the following three forms:

```
\special{overlay filename}   % absolute positioning

\special{include filename}   % relative positioning

\special{insert filename}    % relative positioning
```

In the first case, the PostScript file to be included will be mapped onto the page at precisely the coordinates it specifies. The page origin (0,0) is at the lower-left corner of the page, y positive upwards, x positive to the right.[2] In the last two cases, the upper-left corner of the bounding box will be placed at the current point. The PostScript file must then contain (usually near the start) a comment of the form:

```
%%BoundingBox: llx lly urx ury
```

specifying the bounding box lower-left and upper-right coordinates in standard PostScript units (big points, equivalent to 1/72 inch). Alternatively, if the comment

```
%%BoundingBox: (atend)
```

is found in the file, the last 1,000 characters of the file will be searched to find a %%BoundingBox comment of the first form. This bounding box information is necessary to establish an appropriate translation and scaling of the plot.

If the PostScript file cannot be opened, or the \special command string cannot be recognized, or for relative positioning, the bounding box cannot be determined, a warning message is issued and the \special command is ignored. Otherwise, the section of the PostScript file between the comment lines

```
%begin(plot)
%end(plot)
```

is copied to the output file surrounded by:

---

[2] The Canon print engine used in laser printers like the Apple LaserWriter is incapable of printing inside a margin of about a half inch from each edge of the page; any plot must compensate for this by biasing coordinates, or introducing a translation matrix.

```
save
% revert to standard 1/72 inch units
300 72 div 300 72 div scale
% if relative positioning then
(xcp(in 1/72in)-11x) (ycp(in 1/72in)-ury) translate
...PostScript file contents...
restore
```

For example, if you specify (while setting output dimensions in your graphics package) a device size of 5 inches for a standard horizontal frame, the bounding box will be declared to be 5 inches wide and $(8.5/11) \times 5 = 3.86$ inches high, so a TeX manuscript requiring the graphic could have the following commands at the start of a new paragraph:

```
\special{include graphicfilename}
\vspace*{4.5in}
```

As mentioned earlier a graph is often best displayed in a **figure** environment. You can put the caption above the figure by typing:

```
\begin{figure}[h]
  \caption{This caption is above the plot.}
  \special{include plot.ps}
  \vspace*{4.5in}
\end{figure}
```

or below it by typing:

```
\begin{figure}[h]
  \special{include plot.ps}
  \vspace*{4.5in}
  \caption{This caption is below the plot.}
\end{figure}
```

Note the positioning of the \vspace* command—it *follows* the \special command so that sufficient space is left for the graphic in the figure. TeX cannot determine how much space will be needed, since it does not interpret the argument of the \special command; you have to tell it explicitly.

The upper-left corner of the graphic will be placed where the next character would be typeset, even if that occurs in the middle of a line. Since that is unlikely to be what you want, you should make sure that the **figure** environment is preceded by a blank line, so that it looks like a separate paragraph. Indentation from the left margin can be achieved by use of the \hspace* command, and vertical positioning by the \vspace* command.

More precise electronic placement of PostScript images can be achieved with a combination of Arbortext Corp.'s DVIPS PostScript printer driver for IBM-compatible PCs, and a public-domain software program developed by Trevor Darrell at Pennsylvania State University called **psfig**.

Problems

Problem 14.1 Redefine the \begin{quotation} and \end{quotation} commands with \bq and \eq.

Problem 14.2 Create a custom environment such that quotations are automatically italicized.

Problem 14.3 Create a custom **newtheorem** environment called **post** such that each item is called "Postulate."

Problem 14.4 How would you change pagination to force a given **page** counter to a value of 107?

Problem 14.5 How would you reserve space in a document to later paste in a 3.5-inch-tall graphic, and prelabel the space with a figure caption that says, "NASA's Projected Space Station Costs, 1989–2000"? The caption should be on the bottom of the figure.

Deciphering Error Messages

As you begin processing LaTeX input files, you are bound to encounter error messages. Errors generally result from two sources: either (1) you mistype commands, or (2) misuse commands in describing your document's structure. This chapter will introduce the art of deciphering error messages and learning how to fix them.

15.1 Common Errors

Beginning LaTeXers and veterans alike commit the following types of input errors:

1. Forgetting to type a \ character before the seven special LaTeX symbols: # $ % & _ { }

2. Omitting a matching curly brace to complete a specified environment.

3. Misspelling environment or command names.

4. Using symbols such as the ^ sign in normal text that can only be used in math mode.

5. Accidentally entering math mode by typing a $ sign without prefacing it with a \ symbol.

6. Forgetting to correctly match formula delimiters—e.g. closing the displaymath environment with \) instead of \].

7. Omitting one of these commands:

   ```
   \documentstyle
   \begin{document}
   \end{document}
   ```

15.2   Interpreting Error Messages

Error messages produced while processing input files through LaTeX can be intimidating. Most of them contain a lot of information that is irrelevant to fixing errors. In addition to pinpointing the line number of your input file that contains the error, most messages are filled with technical details of interest only to people skilled in programming native LaTeX and TeX commands. Since this part is of little use to new users, you should ignore it.

Many errors are caused by the mistakes outlined in Section 15.1. A little logic and patience will quickly get you back on track. Error messages usually have the same basic form. For example, if, on line 54 of an input file, you mistakenly typed \begin{centr} (omitting the "e"), you would see the following error message while processing the file through LaTeX:

```
LaTeX error.  See LaTeX manual for explanation.
              Type H <return> for immediate help.
! Environment centr undefined.
\@latexerr ...for immediate help.}\errmessage {#1}
1.54 \begin{centr}

?
```

Here's how to interpret this message:

1. LaTeX's 24 error messages are described in Section 15.3.

2. If you were to type "H" at the "?" prompt, you could enter interactive commands to try to recover from the error. We'll come back to this in a moment.

3. The line beginning with the ! symbol tells you exactly what the error was—in this case, the "centr" environment doesn't exist, which is why it is undefined.

4. The next line gives technical LaTeX information, which you can ignore. The next line tells you the line number of your input file that contains the offending command, and indicates the exact error. It is not important, but it will save you time, if you use a text editor or word processor that can move your cursor to a designated line number.

5. The question mark prompts for your interactive response to the error message.

If you were to type "H" for "help," you would see a message like this:

```
Your command was ignored.
Type I <command> <return>  to replace it with another command,
or  <return>  to continue without it.
?
```

If you don't want to interactively type fixes at this point (none of which are recorded in your input file anyway), sometimes you can bully your way through errors by pressing the return key and get a final .dvi file. Naturally the previewed or printed results will show the effects of your error. This usually is the best approach. All errors are recorded in a text file having the same name as your input file, but ending in .log or .lis. You can go back and read these messages later as you compare the source file with the printed output file.

Sometimes pressing the return key won't yield any forward success. In these cases, you must do something else. You may be tempted to immediately type "E" to leave the LaTeX processing session and make the required change. Since beginners usually make lots of mistakes, this is not the best option. The preferred approach is to type "R" for "run without stopping"; this records all succeeding errors in the .log file. You may then print this file as a reference to work out errors in the input file.

Occasionally you may encounter a TeX error instead of a LaTeX error. An explanation of TeX error messages is beyond the scope of this book. Should this happen, see [4, Chapter 27] for detailed explanations.

15.3 LaTeX Error Messages

Following is a complete list of LaTeX error messages:

! Bad \line or \vector argument. A picture environment error caused by an improper first argument in a \line or \vector command, used to determine the slope.

! Bad math environment delimiter. Caused when you use a math mode start command (\[or \() while in math mode; same error occurs when math mode end command is issued while in standard text mode.

! Bad use of \\. You can't use this command between paragraphs to force an extra line skip. Use the \vskip command instead.

! \begin{...} ended by \end{...}. This error happens when you end an environment with a different \begin command. It typically happens when you misspell the environment name.

!   `Can be used only in preamble.` The commands, \documentstyle, \includeonly, \makeindex, \makeglossary, and \nofiles can only be used before the \begin{document} command.

!   `Command name ... already used.` Names defined by \newcommand, \newenvironment, \newlength, \newsavebox, or \newtheorem can only be defined once—i.e., you can't define the same name two different ways. The same applies to counters defined with \newcounter.

!   `Counter too large.` When you number objects with letters, you can't go higher than a value of 26.

!   `Environment ... undefined.` Error occurs when you are trying to use an environment that is not defined. Check to make sure that you did not misspell a proper environment name.

!   `Float(s) lost.` An output error that causes figures, marginal notes, or tables to be lost. Caused by **figure** or **table** environment or \marginpar command used inside a **minipage** or \parbox.

!   `Illegal character in array arg.` An incorrect character was used in an **array** or **tabular** environment, or in a **multicolumn**'s second argument.

!   `Missing \begin{document}.` You need this to start processing a document. Check to see if you misspelled something in the preamble or left out \begin{document}.

!   `Missing p-arg in array arg.` An **array** or **tabular** environment or a \multicolumn's second argument must have a command statement surrounded by curly braces that follows any "p" used in the setup expression.

!   `Missing @-exp in array arg.` An **array** or **tabular** environment or a \multicolumn's second argument must have a command statement that follows any "@" used in the setup expression.

!   `No such counter.` You forgot to create a new counter before issuing a \addtocounter or \setcounter command, or misspelled the counter name. Make sure the \newcounter command was typed within the preamble.

!   `Not in outer par mode.` Don't use **figure** or **table** environments in math mode or parboxes.

! `\pushtabs` and `\poptabs` don't match. Make sure these commands match, especially before you end a **tabbing** environment.

! Something's wrong--perhaps a missing `\item`. LaTeX is confused by something strange. This suggestion is that you forgot an `\item` command in a list environment.

! Tab overflow. You issued a `\=` command in the **tabbing** environment that exceeded the maximum number of allowable tab stops.

! There's no line here to end. You tried to create space between paragraphs with a `\\` command when you should have used a `\vspace` command.

! This may be a LaTeX bug. On rare occasions, LaTeX has been known to have minor, obscure bugs. It's just your luck that you found one.

! Too deeply nested. LaTeX normally permits four levels of nesting list environments; apparently you tried to go too deep.

! Too many unprocessed floats. A likely error when you have lots of **figure** and **table** environments. LaTeX can keep track of a limited number while finishing page processing before it runs out of memory. The error often is triggered when you have a large environment that won't fit on one page, or when you haven't specified an optional placement argument. Reduce its size or use the p option to try to solve the error.

! Undefined tab position. You forgot to define a tab position with the `\=` command in the **tabbing** environment.

! `\<` in mid line. The `\<` command can be used only at the start of a line; you used it somewhere in the middle of a line within the **tabbing** environment.

15.4   LaTeX and TEX Warnings

Sometimes you will see LaTeX or TEX warning messages while processing your input file. Usually these fall into two categories: (1) undefined citations, labels, and references, and (2) overfull or underfull `\hbox` or `\vbox` errors.

Undefined citations, labels, and references generally happen when the actual reference lies further along in the document. When you process the input file through LaTeX a second time, the correct definitions are picked up from the .aux file created during the first pass. If you see an undefined reference during the second pass, you probably forgot to define it in the input file.

Overfull or underfull boxes (see Section 3.5) are a little trickier to handle. If you encounter an overfull \hbox, it means that the last word on the indicated line was unable to hyphenate itself in an optimal position; hence, the word extended past the right margin by an indicated number of printer's points. There are 72.27 points to an inch. If the distance is under about five points, you probably won't notice the difference. Typically this problem results because LaTeX doesn't know where to correctly hyphenate the last word on the line. Try placing the \- sign in these words to tell LaTeX where it can perform troublesome hyphenations. Sometimes you must alter the text (move words around, select new words, etc.) to solve an overfull \hbox problem.

An overfull \vbox means LaTeX couldn't find a good place to issue a page break. Look at the finished page to decide if you must manually improve its appearance. If it needs help, you might try forcing a page break with the \pagebreak command where you think the break should occur. A \newpage command forces a page break at the exact spot where you type the command. The underfull \vbox means you have extra vertical space on that page. This error usually is caused either by a misused \\ (line break) or \newline command.

15.5   More Help on Pinpointing Errors

If you have errors in a multipage LaTeX document, pay attention to where they occur with respect to the on-screen page number indicators. Each time you see a page number enclosed in square brackets, LaTeX is telling you that that page has finished processing. For example, if [24] appears on the screen followed by an error, the error has occurred on page 25 of your final document.

Complicated errors might require you to eliminate portions of your input file until you find the offending section. This is easily done by copying your source file into a temporary file, erasing the obviously good parts, and processing the balance through LaTeX. Repeat this until you isolate the error.

As a last resort, look for suggestions in [6] and your *Local Guide*, or from your LaTeX support person. As you gain experience, errors will become less frequent. When they do occur, you'll be able to handle them more easily.

The rewards of doing your own typesetting far outweigh the occasional problem of dealing with processing errors.

Answers to Chapter Exercises

Problem 3.1

This input file should look like this:

```
\documentstyle{article}
\begin{document}

Vigorous writing is concise.  A sentence should contain
no unnecessary words, a paragraph no unnecessary
sentences, for the same reason that a drawing should
have no unnecessary lines and a machine no unnecessary
parts.  This requires not that the writer make all his
sentences short, or that he avoid all detail and treat
his subjects only in outline, but that every word tell.

\end{document}
```

Problem 3.2

Optional document styles are used by replacing **article** designation with one of these choices:

```
\documentstyle{report}
\documentstyle{book}
\documentstyle{letter}
```

Only one document style can be used at a time in your document. Your *Local Guide* or LaTeX administrator will tell you if your site has created document styles customized for special purposes (e.g. dissertation or thesis format).

Problem 3.3

```
The \$ sign is used to start or end the creation of
simple mathematical symbols or expressions. The \&
sign separates columns in tables. Non-printing,
in-text notes are preceded by the \% sign. Macros
often contain designated parameters signified by the
\# character. Subscripts are created with the \_
symbol. When you wish to segregate a block of characters
to be treated as one unit, you surround them with the
\{ and \} delimiters.
```

Problem 3.4

```
``Clusters of stars---for example, the Pleiades---are
chosen because we can assume that such stellar
aggregations are approximately coeval.\ . \ . \ .  In
Figures 6--5 and 6--6, we see spectrum-luminosity
diagrams for two different stellar clusters.''
```

Note how the \ character is used to make proper spacing between the periods to create a four-dot ellipsis.

Problem 3.5

```
Il y eut \'{a} toute les \'{e}poques des enfants qui
apprirent \'{a} lire, \'{a} \'{e}crire, \'{a} computer;
des jeunes gens qui, comme ma s{\oe}ur, suivirent des
cours de lettres et de sciences.
```

Problem 3.6

End-of-sentence spacing is prevented by typing a \ character that immediately follows the period, like this:

```
Capt.\ Roberts
```

Problem 3.7

The \vspace command is used to create "vertical space" between paragraphs like this:

```
. . . last line of the top paragraph.

\vspace{5in}

First line of the bottom paragraph. . .
```

Problem 3.8

The \\ command forces a mid-paragraph line-break. In this example, a new line would appear after the word "paragraph," and the word "lines."

```
Vigorous writing is concise.  A sentence should contain
no unnecessary words, a paragraph \\
no unnecessary sentences, for the same reason that a
drawing should have no unnecessary lines \\
and a machine no unnecessary parts.
```

Problem 3.9

Use the in-text "comment" command (the % sign). Everything on a line that follows a % sign is ignored by TeX.

```
% Version 3.2
% Last Date Modified: October 3, 1994
% Author: Eric J. Sampson

\documentstyle{article}
\begin{document}
```

If you want the percent sign to appear in the document (and all text that follows it), simply type a \ before the % sign.

Problem 4.1

```
\begin{center}
{\it Mathematics Into Type: Copy Editing and
Proofreading of Mathematics for Editorial
Assistants and Authors}\\
by \\
Ellen Swanson
\end{center}
```

Problem 4.2

```
\begin{flushright}
Professor Michael Daniels \\
Department of Economics \\
231 Sampson Hall \\
Illinois State University \\
Chicago, IL  60611
\end{flushright}
```

Problem 4.3

```
\begin{flushleft}
The earliest, and still the most fundamental, astronomical
distance determination involves triangulation, the same
method used by surveyors to determine the distance to an
inaccessible point.  The astronomer observes the star of
interest from two different, widely separated points, and
notes the apparent motion of the star against a background
of more distant objects.
\end{flushleft}
```

Problem 4.4

```
\begin{itemize}
    \item Star
    \item Hierarchical Tree
    \item Loop
    \item Bus
    \item Ring
    \item Web
\end{itemize}
```

Problem 4.5

```
\begin{enumerate}
    \item Star
    \item Hierarchical Tree
    \item Loop
    \item Bus
    \item Ring
    \item Web
\end{enumerate}
```

Problem 4.6

```
\begin{description}

\item[Star:] All traffic routed to and handled by a central
computer.  This topology looks like a spoked wheel, with
the central computer at the hub and recipient computers
at the end of each spoke.

\item[Hierarchical Tree:] Typical to mainframe environments,
this topology looks like an upside-down tree.  Top computer
coordinates network; intermediate computers control traffic
at their level and below.

\item[Loop:] Typical to workgroups, this topology is a
daisy-chain of computers that form a ring.  Each computer
must be capable of performing all network communications
functions.

\item[Bus:] A network backbone where all computers share a
common communications line (bus).  This bus is not joined
in a loop, and essentially forms a straight line.

\item[Ring:] A cross between a loop and bus topology.  Failed
nodes, however, do not cause the network to stop working
because they attach off the main bus.

\item[Web:] A spaghetti-like topology where each node is
attached via dedicated links.

\end{description}
```

Problem 4.7

Use the quotation environment to duplicate the problem. If you don't want the first line in each paragraph to be indented, use the quote environment.

```
\begin{quotation}
Another noteworthy characteristic of this manual is
that it doesn't always tell the truth.  When certain
concepts of \TeX\ are introduced informally, general
rules will be stated; afterwards you will find that
```

```
the rules aren't strictly true.  In general, the later
chapters contain more reliable information than the
earlier ones do. The author feels that this technique
of deliberate lying will actually make it easier for
you to learn the ideas. Once you understand a simple
but false rule, it will not be hard to supplement that
rule with its exceptions.
\end{quotation}
```

Problem 4.8

```
\begin{center}
{\bf H.D. (Hilda Doolittle)

Scribe}
\end{center}
\begin{verse}
Wildly dissimilar \\
yet actuated by the same fear \\
the hippopotamus and the wild-deer \\
hide by the same river.

Strangely disparate \\
yet compelled by the same hunger, \\
the cobra and the turtle-dove \\
meet in the palm-grove.

{\footnotesize 1920s}
\end{verse}
```

Problem 4.9

```
\begin{verbatim}

The first command in a \LaTeX\ input file
is this:

    \documentstyle{article}

An input file's text is surrounded by these commands:

    \begin{document}

    \end{document}
```

`\end{verbatim}`

Problem 5.1

To create **boldface** type, you must use the \bf command in one of two ways. One way is to preface a string of characters you wish to boldface with \bf. When you want to switch back to roman type, you type \rm. Here is an example:

```
This text appears in roman (normal) type, while \bf this
text appears in boldface type\rm.  Everything after the
last command switches back to roman type.
```

Alternatively, you may encapsulate the boldface type like this:

```
This text appears in roman (normal) type, while {\bf this
text appears in boldface type}.  Only the text within the
curly brace delimiters is boldfaced.
```

Problem 5.2

```
{\tiny This type is tiny.}

{\scriptsize This type is scriptsize.}

{\footnotesize This type is footnotesize.}

{\small This type is small.}

{\normalsize This type is normalsize.}

{\large This type is large.}

{\Large This type is Large.}

{\LARGE This type is LARGE.}

{\huge This type is huge.}

{\Huge This type is Huge.}
```

Problem 5.3

```
Almost every kid knows that bee stings
are {\Large\bf very painful}.
```

Problem 5.4

```
\newcommand{\til}{\char '176}
```

Problem 6.1

```
\[
\sum_{n=1}^{100}x_n
\]
```

Problem 6.2

```
\[
\left\lfloor
\frac{x^{2k-2}}{\sum_{i=0}^{k-1}a_ix^i}
\right\rfloor
\]
```

Problem 6.3

```
\[
bn\sum_{i=0}^{\log_c n} r^i = bn
\frac{r^{1+\log_c n}-1}{r-1}
\]
```

Problem 6.4

```
\[
\sum_{i=2}^{n-1}i\log_e i \leq \int_2^n x
\log_e x dx \leq
\frac{n^2 \log_e n}{2} - \frac{n^2}{4}
\]
```

Problem 6.5

```
\[
\sum_{i=0}^{n/2-1} (a^2)^i = \prod_{i=o}^{k-2}
[1 + (a^2)^{2^{i}}]
= \prod_{i=1}^{k-1} [1 + a^{2^{i}}]
\]
```

Problem 7.1

```
\[ M= \left[
\begin{array}{rrrr}
0 & 0 & 1 & 2 \\
0 & 0 & 3 & 0 \\
1 & -1 & 0 & 1 \\
2 & 0 & -1 & 3
\end{array}
\right] \]
```

Problem 7.2

```
\[
\left[
\begin{array}{rrrr}
1 & 0 & 0 & 0 \\
0 & 0 & 1 & 1 \\
1 & 0 & 0 & 1 \\
0 & 0 & 1 & 0
\end{array}
\right]
\left[
\begin{array}{rrrr}
0 & 1 & 0 & 0 \\
1 & 1 & 0 & 0 \\
1 & 0 & 0 & 0 \\
0 & 0 & 0 & 1
\end{array}
\right]
\]
```

Problem 7.3

```
\[
\begin{array}{r|rrr}
i & a_i & x_i & y_i \\
\hline
0 & 57 & 1 & 0 \\
1 & 33 & 0 & 1 \\
2 & 24 & 1 & -1 \\
3 & 9 & -1 & 2 \\
4 & 6 & 3 & -5 \\
5 & 3 & -4 & 7
\end{array}
\]
```

Problem 7.4

```
\[
\begin{array}{c|c|c|c|c|c|c|c|}
  & a & b & e & i & t & \$ \\
\hline
S & S \rightarrow a & & & S \rightarrow iCtSS' & & \\
\hline
  & & & S' \rightarrow \epsilon & & & \\
S' & & & S' \rightarrow eS & & & S' \rightarrow \epsilon \\
\hline
C & & C \rightarrow b & & & & \\
\hline
\end{array}
\]
```

Problem 7.5

```
\[
T(n) = \left\{ \begin{array}{ll}
              O(n),              &  {\rm if} a<c, \\
              O(n \log n),       &  {\rm if} a=c, \\
              O(n^{\log_c{a}}),  &  {\rm if} a>c.
              \end{array}
       \right.
\]
```

Problem 7.6

```
\begin{eqnarray*}
T(m) & \leq &
\frac{en}{4m}\left[4M \left(\frac{m}{2}\right) + 4^2M
\left(\frac{m}{2^2}\right) + \cdots + 4^{\log
m}M(1)\right]+bnm \\
& \leq & \frac{en}{4m}\sum_{i=1}^{\log m} 4^iM
\left(\frac{m}{2^i}\right) + bnm.
\end{eqnarray*}
```

Problem 7.7

```
\begin{eqnarray*}
\left(\sum_{i=0}^{n-1}a_ix^i\right)
\left(\sum_{j=0}^{n-1}b_jx^u\right)
= \sum_{k=0}^{2n-2}c_kx^k,
& {\rm where} &
c_k = \sum_{m=0}^{n-1}a_mb_{k-m}.
\end{eqnarray*}
```

Problem 8.1

The first method is like this:

```
\begin{tabbing}
EIA Designation \= CCITT Designation \= Name \\
AA \> 101 \> Protective Ground \\
AB \> 102 \> Signal Ground \\
BA \> 103 \> Transmitted Data \\
BB \> 104 \> Received Data \\
CA \> 105 \> Request to Send \\
CB \> 106 \> Clear to Send \\
CC \> 107 \> Data Set Ready \\
CF \> 109 \> Data Channel Received Line Signal Detector
\end{tabbing}
```

The second method requires space reservation between columns like this:

```
\begin{tabbing}
xxxxxxxxxxxxxxxx\=xxxxxxxxxxxxxxxxx\= \kill
EIA Designation \= CCITT Designation \= Name \\
AA \> 101 \> Protective Ground \\
AB \> 102 \> Signal Ground \\
BA \> 103 \> Transmitted Data \\
BB \> 104 \> Received Data \\
CA \> 105 \> Request to Send \\
CB \> 106 \> Clear to Send \\
CC \> 107 \> Data Set Ready \\
CF \> 109 \> Data Channel Received Line Signal Detector
\end{tabbing}
```

The final method requires that you type column widths like this:

```
\begin{tabbing}
\hspace{1in}\=\hspace{1in}\=\hspace{3in} \kill
EIA Designation \= CCITT Designation \= Name \\
AA \> 101 \> Protective Ground \\
AB \> 102 \> Signal Ground \\
BA \> 103 \> Transmitted Data \\
BB \> 104 \> Received Data \\
CA \> 105 \> Request to Send \\
CB \> 106 \> Clear to Send \\
CC \> 107 \> Data Set Ready \\
CF \> 109 \> Data Channel Received Line Signal Detector
\end{tabbing}
```

Problem 8.2

```
\begin{tabular}{|c|c|c|} \hline
EIA Designation & CCITT Designation & Name \\ \hline
AA    & 101  & Protective Ground         \\ \hline
AB    & 102  & Signal Ground             \\ \hline
BA    & 103  & Transmitted Data          \\ \hline
BB    & 104  & Received Data             \\ \hline
CA    & 105  & Request to Send           \\ \hline
CB    & 106  & Clear to Send             \\ \hline
CC    & 107  & Data Set Ready            \\ \hline
CF    & 109  & DC Rec'd Line Signal Detec. \\ \hline
\end{tabular}
```

Problem 8.3

```
\begin{table}[h]
\caption{Low-Speed Asynchronous Full-Duplex
         Modem Interface Leads}
\begin{tabular}{|c|c|c|} \hline
EIA Designation & CCITT Designation & Name \\ \hline
AA    & 101  & Protective Ground         \\ \hline
AB    & 102  & Signal Ground             \\ \hline
BA    & 103  & Transmitted Data          \\ \hline
BB    & 104  & Received Data             \\ \hline
CA    & 105  & Request to Send           \\ \hline
CB    & 106  & Clear to Send             \\ \hline
CC    & 107  & Data Set Ready            \\ \hline
CF    & 109  & DC Rec'd Line Signal Detec. \\ \hline
\end{tabular}
\end{table}
```

Problem 10.1

```
\documentstyle[12pt,twocolumn]{article}
```

Problem 10.2

Type the following in your document preamble:

```
\pagestyle{myheadings}
\markboth{Great Research Projects of the 1980s}{Psychosis
          and Typesetting}
```

Problem 10.3
Type the following in your document preamble:

```
\textwidth 5in
\textheight 7in
\topmargin 2in
```


Problem 11.1

```
@BOOK{brooks,
    KEY = "america",
    AUTHOR = "   ",
    TITLE = "America---20th Century Poetry: Landscapes
            of the Mind",
    ADDRESS = "Evanston, Ill.",
    PUBLISHER = "McDougal, Littell \& Co.",
    YEAR = "1973"
    }
@BOOK{frost,
    EDITOR = "Edward Connery Lathem",
    TITLE = "The Poetry of Robert Frost",
    ADDRESS = "New York",
    PUBLISHER = "{Holt, Rinehart and Winston}",
    YEAR = "1969"
    }
@BOOK{spiker,
    AUTHOR = "Sina Spiker",
    TITLE = "Indexing Your Book: A Practical Guide for Authors",
    ADDRESS = "{Madison, Wisc.}",
    PUBLISHER = "Univ. of Wisc. Press",
    YEAR = "1954"
    }
```

Problem 11.2
References would be cited in text as follows:

```
\cite{brooks}
\cite{frost}
\cite{spiker}
```

Problem 11.3

```
@BOOK{style,
    AUTHOR = "{Strunk, Jr.}, William and White, {E. B.}",
    TITLE = "The Elements of Style",
    ADDRESS = "New York",
    PUBLISHER = "The Macmillan Co.",
    YEAR = "1972"
    }
```

Problem 12.1

```
Vigorous writing is concise.  A sentence\index{sentence}
should contain no unnecessary words, a paragraph
\index{paragraph} no unnecessary sentences, for
the same reason that a drawing should have no unnecessary
lines and a machine no unnecessary parts.  This requires
not that the writer\index{writer} make all his sentences
short, or that he avoid all detail and treat his subjects
only in outline, but that every word tell.
```

Problem 12.2

```
Vigorous writing\index{writing!vigorous}
```

Problem 12.3

```
Vigorous writing\index{writing!vigorous|see{Chapter~12}}
```

Problem 12.4

To mark the range for indexing, you would type the following where the "vigorous writing" discussion begins:

```
\index{writing!vigorous|(}
```

You would then type the following where the "vigorous writing" discussion ends:

```
\index{writing!vigorous|)}
```

Problem 13.1

You type the following command in the preamble:

```
\documentstyle[twocolumn]{article}
```

Note that the twocolumn command is a document style option; other options can be included, each of which is separated by a comma. The article document style can be substituted with another choice.

Problem 13.2

```
\documentstyle[proc]{article}
```

Problem 13.3

You type the \twocolumn command to switch to two-column mode in the middle of your text's body. This command automatically starts a new page on which the two-column mode begins. You type \onecolumn to revert to single-column mode.

Problem 13.4

```
\begin{center}
\begin{minipage}{3in}

Beginning \LaTeX\/ers would do well to take counsel
from the ''Grand Wizard of \TeX,'' Donald
Knuth:\footnote{Cited from {\it The \TeX\/book}, p. vi.}

\begin{quote}
When you first try to use \TeX, you'll find that some
parts of it are very easy, while other things will take
some getting used to.  A day or so later, after you have
successfully typeset a few pages, you'll be a different
person; the concepts that used to bother you will now
seem natural, and you'll be able to picture the final
result in your mind before it comes out of the machine.
\end{quote}

\end{minipage}
\end{center}
```

Problem 14.1

Type the following commands in the preamble:

```
\newcommand{\bq}{\begin{quotation}}
\newcommand{\eq}{\end{quotation}}
```

Problem 14.2

Use the \newenvironment command as follows:

```
\newenvironment{myquote}{\begin{quote} \em}{\end{quote}}
```

To start this environment, you type \begin{myquote}; to end it you type \end{myquote}.

Problem 14.3

```
\newtheorem{post}{Postulate}
```

To start this environment, you type \begin{post}; to end it you type \end{post}.

Problem 14.4

Type \setcounter{page}{107} where you want this page number to oc-cur. Subsequent page numbers will be increased from this value unless you change it again.

Problem 14.5

Type the following commands where you want to place the graphic:

```
\begin{figure}
   \vspace{3.5in}
   \caption{NASA's Projected Space Station
           Costs, 1989--2000}
\end{figure}
```

Sample Input Files

The following sample input files illustrate some LaTeX features. The left page shows the final output, and the right page show the input which produced it. Try experimenting with this input and use it to create your own documents.

Typing Text

Text is entered as you usually do with a word processor or text editor. A few characters, however, are reserved for initiating LaTeX commands. These include:

$$\# \quad \$ \quad \% \quad \& \quad ~ \quad \_ \quad \char`\^ \quad \backslash \quad \{ \quad \}$$

Except for the ~ _ \ characters, all of these symbols can be created by prefacing them with a \ character (e.g. \#). LaTeX ignores extra spaces between words. You can create an extra interword space by adding the \ character between two words. A \\ sequence will force a line break in the middle of a paragraph. Paragraphs are separated by one or more blank lines. Quote marks are created by typing ' ' and ' ' instead of using the " character. Hyphens come in three types: single dashes to link compound words (e.g. hyper-active), double dashes for number ranges (e.g. 1984–87), and triple dashes to connect compound phrases (e.g. she was there—four, maybe five years ago).

Type Styles and Sizes

Typeset documents normally do not have underlined text; instead these words are *emphasized* in italic type. The italics were created by typing {\em emphasized}. The \em switched type to emphasized mode; the curly braces surrounding the word were used as a grouping technique to restrict emphasis to text only within their boundaries. Other options allow you to have typefaces like **boldface** and sans serif, as well as change type sizes ranging from this to this.

```
\documentstyle{article}          % chose the "article" document style
\begin{document}                 % this starts the document
\section{Typing Text}            % creating a section
```

Text is entered as you usually do with a word processor or
text editor. A few characters, however, are reserved for initiating
\LaTeX\ commands. These include:

```
\begin{center}                   % creating a "centering" environment
\verb|# $ % & ~ _ ^ \ { }|
\end{center}
```

Except for the \verb|~ _ \| characters, all of these symbols can be
created by prefacing them with a \verb|\| character (e.g.\ \verb|\#|).
\LaTeX\ ignores extra spaces between words. You can create an extra
interword space by adding the \verb|\| character between two words.
A \verb|\\| sequence will force a line break in the middle of a
paragraph. Paragraphs are separated by one or more blank lines.
Quote marks are created by typing \verb|``| and \verb|''| instead of
using the \verb|"| character. Hyphens come in three types: single
dashes to link compound words (e.g.\ hyper-active), double dashes for
number ranges (e.g.\ 1984--87), and triple dashes to connect compound
phrases (e.g.\ she was there---four, maybe five years ago).

```
\subsection{Type Styles and Sizes}  % creating a sub-section
```

Typeset documents normally do not have \underline{underlined} text;
instead these words are {\em emphasized\/} in italic type. The
italics were created by typing \verb|{\em emphasized}|. The \verb|\em|
switched type to emphasized mode; the curly braces surrounding the
word were used as a grouping technique to restrict emphasis to text
only within their boundaries. Other options allow you to have
typefaces like {\bf boldface} and {\sf sans serif}, as well as change
type sizes ranging {\tiny from this} {\Huge to this}.

```
\end{document}
```

1. (10) Find the equation of a line passing through the x-axis at 3 and
   parallel to the line tangent to the curve

$$y = 3x^3 - 2x + 5$$

at $x = 2$. Transform the equation of the desired line into slope-
intercept form.

2. (15) If $g(y) = 2^{3y}$, then is $g(u)g(v) = g(uv)$ or $g(u)g(v) = g(u+v)$,
   or are they both false? Give reasons for your answer.

3. (20) Perform the following integrations:
   (a)   (b)

$$\int \frac{\sin \sqrt{x}\,dx}{\sqrt{x}}$$

$$\int x^2 dy$$

4. (20) Use L'Hôpital's rule to evaluate the following limits:
   (a)   (b)

$$\lim_{x \to 0} \frac{x \sin 3x}{5x^2}$$

$$\lim_{x \to 0} \frac{1}{x} - \frac{1}{\sqrt{x}}$$

65 points total

```
\documentstyle[12pt]{article}
\begin{document}
\noindent Math 11.09 (12:10 noon)---Fall, 1987 \hfill Quiz \\
Professor John Smith \hfill November 9, 1987

\begin{enumerate}

\item (10) Find the equation of a line passing through the
x-axis at $3$ and parallel to the line tangent to the curve
\[ y=3x^{3}-2x+5 \] at $x=2$.  Transform the equation of the
desired line into slope-intercept form.

\item (15) If $g(y)=2^{3y}$, then is $g(u)g(v)=g(uv)$ or
$g(u)g(v)=g(u+v)$, or are they both false?  Give reasons for
your answer.

\item (20) Perform the following integrations:
    \begin{enumerate}                 % begins numbered list environment
    \begin{minipage}[t]{60mm}    % minipage used to create 2-columns
    \item \[ \int \frac{\sin \sqrt{x} dx}{\sqrt{x}} \]
    \end{minipage}
    \begin{minipage}[t]{60mm}    % minipage used to create 2-columns
    \item \[ \int x^{2}dy \]
    \end{minipage}
    \end{enumerate}

\item (20) Use L'H\^{o}pital's rule to evaluate the following limits:
    \begin{enumerate}
    \begin{minipage}[t]{60mm}
    \item \[ \lim_{x\rightarrow 0}
            \frac{x \sin 3x}{5x^2}     \]
    \end{minipage}
    \begin{minipage}[t]{60mm}
    \item \[ \lim_{x\rightarrow 0}
            \frac{1}{x} - \frac{1}{\sqrt{x}}    \]
    \end{minipage}
    \end{enumerate}
\end{enumerate}
\noindent 65 points total

\end{document}
```

[Sample input from a technical paper[1]]

The constant π was computed using Borweins' quartically convergent algorithm, which was discovered in 1985 [5]. This algorithm is as follows: Let $a_0 = 6 - 4\sqrt{2}$ and $y_0 = \sqrt{2} - 1$. Iterate

$$
\begin{aligned}
y_{k+1} &= \frac{1 - (1 - y_k^4)^{1/4}}{1 + (1 - y_k^4)^{1/4}} \\
a_{k+1} &= a_k(1 + y_{k+1})^4 - 2^{2k+3} y_{k+1}(1 + y_{k+1} + y_{k+1}^2)
\end{aligned}
$$

Then a_k converges quartically to $1/\pi$: each successive iteration approximately *quadruples* the number of correct digits in the result.

Euler's constant γ was calculated using the following formulas, which are an improvement of a technique previously used by Sweeney [10].

$$
\begin{aligned}
\gamma &= \frac{2^n}{e^{2^n}} \sum_{m=0}^{\infty} \frac{2^{nm}}{(m+1)!} \sum_{t=0}^{m} \frac{1}{t+1} - n\log 2 + O(\frac{1}{2^n e^{2^n}}) \\
\log 2 &= \sum_{k=1}^{\infty} \frac{1}{(2k-1)3^{2k-1}}
\end{aligned}
$$

---

[1] This sample, as well as Table B.1, are cited courtesy of David H. Bailey.

```latex
\documentstyle{article}
\begin{document}
\thispagestyle{empty}        % omits pagination on this page

The constant $\pi$ was computed using Borweins' quartically
convergent algorithm, which was discovered in
1985 \cite{yourbibref}.  This algorithm is as follows:

Let $a_0 = 6 - 4 \sqrt{2}$ and $y_0 = \sqrt{2} - 1$.  Iterate

\begin{eqnarray*}
y_{k+1} &=& \frac{1 - (1 - y_k^4)^{1/4}}{1 +
    (1 - y_k^4)^{1/4}} \\
a_{k+1} &=& a_k(1 + y_{k+1})^4 - 2^{2k+3}y_{k+1}(1 +
    y_{k+1} + y_{k+1}^2)
\end{eqnarray*}

\noindent
Then $a_k$ converges quartically to $1/\pi$: each successive
iteration approximately {\it quadruples\/} the number of
correct digits in the result.

Euler's constant $\gamma$ was calculated using the following
formulas, which are an improvement of a technique previously
used by Sweeney [10].

\begin{eqnarray*}
\gamma &=& \frac{2^n}{e^{2^n}} \sum_{m=0}^{\infty}
    \frac{2^{nm}}{(m+1)!}\sum_{t=0}^{m} \frac{1}{t+1}
    \; - \; n \log 2 \; + \; O(\frac{1}{2^n
    e^{2^n}}) \\
\log 2 &=& \sum_{k=1}^{\infty} \frac{1}{(2k - 1) 3^{2k-1}}
\end{eqnarray*}

\end{document}
```

Food Menu

Spanish Rice:

Cook until yellow in 4 tablespoons of margarine:

    1   Chopped onion
    1   Green pepper
    $\frac{1}{4}$ Cup diced celery

Add and cook slowly 15 minutes:

    1   Can peeled tomatoes
    1   Teaspoon salt

Gently stir in hot, drained, boiled rice (1 cup uncooked) and cook 10 minutes. Serve hot; sprinkle with grated cheese.

```
% Sample food menu document
%
\documentstyle{article}
\begin{document}
\thispagestyle{empty}          % omits pagination on this page
\begin{center}{\LARGE\bf Food Menu}\end{center}

\noindent {\large\bf Spanish Rice}:
\vspace{5mm}                   % adds extra vertical space

Cook until yellow in 4 tablespoons of margarine:

\begin{tabbing}
xxxxxxxxxx\=xxx\=  \kill      % "fixed-width" tabbing method
\> $1$ \> Chopped onion \\    % Since a fraction is used, all #'s
\> $1$ \> Green pepper  \\    % are in math mode for uniformity
\> $\frac{1}{4}$ \> Cup diced celery
\end{tabbing}

Add and cook slowly 15 minutes:

\begin{tabbing}
xxxxxxxxxx\=xxx\=  \kill
\> $1$ \> Can peeled tomatoes \\
\> $1$ \> Teaspoon salt
\end{tabbing}

Gently stir in hot, drained, boiled rice ($1$ cup
uncooked) and cook $10$ minutes.  Serve hot;
sprinkle with grated cheese.

\end{document}
```

Table B.1. Cray-2 Performance Figures

m	CFFT2		CFFTZ			Time Ratio
	CPU Time	Error	CPU Time	Error	MFLOPS	
8	0.00023	5.111×10^{-15}	0.00027	6.078×10^{-15}	37.92	0.846
9	0.00048	5.658×10^{-15}	0.00045	6.130×10^{-15}	51.00	1.061
10	0.00099	6.784×10^{-15}	0.00074	6.913×10^{-15}	69.60	1.347
11	0.00204	1.102×10^{-14}	0.00148	7.052×10^{-15}	76.33	1.380
12	0.00420	1.252×10^{-14}	0.00274	7.608×10^{-15}	89.62	1.531
13	0.00868	3.510×10^{-14}	0.00538	7.865×10^{-15}	98.99	1.614
14	0.01871	5.445×10^{-14}	0.01055	8.430×10^{-15}	108.70	1.773
15	0.03987	2.095×10^{-13}	0.02427	8.555×10^{-15}	101.28	1.643
16	0.07719	2.518×10^{-13}	0.04724	9.092×10^{-15}	10.99	1.634
17	0.16113	6.416×10^{-13}	0.09524	9.248×10^{-15}	116.98	1.692
18	0.32692	9.999×10^{-13}	0.19488	9.758×10^{-15}	121.07	1.678
19	0.67228	7.362×10^{-13}	0.43286	9.847×10^{-15}	115.06	1.553
20	1.39522	4.498×10^{-13}	0.89189	1.035×10^{-14}	117.57	1.564

```
\documentstyle{article}
\begin{document}
\thispagestyle{empty}          % omits pagination on this page
\begin{table}
\caption{Cray-2 Performance Figures}
\begin{tabular}{|r|r|r|r|r|r|r|r|}
\hline
    & \multicolumn{2}{|c|}{CFFT2} & \multicolumn{3}{|c|}{CFFTZ}
    & Time & Error \\
  m & CPU Time & Error & CPU Time & Error & MFLOPS & Ratio
    & Ratio \\
\hline
  8  &  0.00023  &  5.111$\times 10^{-15}$  &  0.00027  &
     6.078$\times 10^{-15}$  &  37.92  &  0.846  &  0.8410 \\
  9  &  0.00048  &  5.658$\times 10^{-15}$  &  0.00045  &
     6.130$\times 10^{-15}$  &  51.00 & 1.061  &  0.9229 \\
 10  &  0.00099  &  6.784$\times 10^{-15}$  &  0.00074  &
     6.913$\times 10^{-15}$  &  69.60  &  1.347  &  0.9813 \\
 11  &  0.00204  &  1.102$\times 10^{-14}$  &  0.00148  &
     7.052$\times 10^{-15}$  &  76.33  &  1.380  &  1.562 \\
 12  &  0.00420  &  1.252$\times 10^{-14}$  &  0.00274  &
     7.608$\times 10^{-15}$  &  89.62 & 1.531  &  1.646 \\
 13  &  0.00868  &  3.510$\times 10^{-14}$  &  0.00538  &
     7.865$\times 10^{-15}$  &  98.99  &  1.614  &  4.462 \\
 14  &  0.01871  &  5.445$\times 10^{-14}$  &  0.01055  &
     8.430$\times 10^{-15}$  &  108.70  &  1.773  &  6.459 \\
 15  &  0.03987  &  2.095$\times 10^{-13}$  &  0.02427  &
     8.555$\times 10^{-15}$  &  101.28  &  1.643  &  24.49 \\
 16  &  0.07719  &  2.518$\times 10^{-13}$  &  0.04724  &
     9.092$\times 10^{-15}$  &  10.99  &  1.634  &  27.69 \\
 17  &  0.16113  &  6.416$\times 10^{-13}$  &  0.09524  &
     9.248$\times 10^{-15}$  &  116.98  &  1.692  &  69.38 \\
 18  &  0.32692  &  9.999$\times 10^{-13}$  &  0.19488  &
     9.758$\times 10^{-15}$  &  121.07  &  1.678  &  102.5 \\
 19  &  0.67228  &  7.362$\times 10^{-13}$  &  0.43286  &
     9.847$\times 10^{-15}$  &  115.06  &  1.553  &  74.76 \\
 20  &  1.39522  &  4.498$\times 10^{-13}$  &  0.89189  &
     1.035$\times 10^{-14}$  &  117.57  &  1.564  &  43.47 \\
\hline
\end{tabular}
\end{table}
\end{document}
```

Mathematical Symbols

Table C.1. Greek Letters

Lowercase					
α	\alpha	ι	\iota	ϱ	\varrho
β	\beta	κ	\kappa	σ	\sigma
γ	\gamma	λ	\lambda	ς	\varsigma
δ	\delta	μ	\mu	τ	\tau
ϵ	\epsilon	ν	\nu	υ	\upsilon
ε	\varepsilon	ξ	\xi	ϕ	\phi
ζ	\zeta	o	o	φ	\varphi
η	\eta	π	\pi	χ	\chi
θ	\theta	ϖ	\varpi	ψ	\psi
ϑ	\vartheta	ρ	\rho	ω	\omega
Uppercase					
A	A	I	I	P	P
B	B	K	K	Σ	\Sigma
Γ	\Gamma	Λ	\Lambda	T	T
Δ	\Delta	M	M	Υ	\Upsilon
E	E	N	N	Φ	\Phi
Z	Z	Ξ	\Xi	χ	\chi
H	H	O	O	Ψ	\Psi
Θ	\Theta	Π	Pi	Ω	\Omega

Table C.2. Binary Operation Symbols

±	\pm	⊓	\sqcap	▷	\rhd
∓	\mp	⊔	\sqcup	⊴	\unlhd
×	\times	∨	\vee	⊵	\unrhd
÷	\div	∧	\wedge	⊕	\oplus
*	\ast	\	\setminus	⊖	\ominus
⋆	\star	≀	\wr	⊗	\otimes
∘	\circ	⋄	\diamond	⊘	\oslash
●	\bullet	△	\bigtriangleup	⊙	\odot
·	\cdot	▽	\bigtriangledown	○	\bigcirc
∩	\cap	◁	\triangleleft	†	\dagger
∪	\cup	▷	\triangleright	‡	\ddagger
⊎	\uplus	◁	\lhd	Ⅱ	\amalg

Table C.3. Relation Symbols

≤	\leq	⪰	\succeq	≈	\approx
≺	\prec	≫	\gg	≅	\cong
⪯	\preceq	⊃	\supset	≠	\neq
≪	\ll	⊇	\supseteq	≐	\doteq
⊂	\subset	⊐	\sqsupset	∝	\propto
⊆	\subseteq	⊒	\sqsupseteq	⊨	\models
⊏	\sqsubset	∋	\ni	⊥	\perp
⊑	\sqsubseteq	⊣	\dashv	|	\mid
∈	\in	≡	\equiv	‖	\parallel
⊢	\vdash	∼	\sim	⋈	\bowtie
≥	\geq	≃	\simeq	⋈	\Join
≻	\succ	≍	\asymp	⌣	\smile
				⌢	\frown

Table C.4. "Log-like" (Always in roman) Function Names

arccos	\arccos	det	\det	ln	\ln
arcsin	\arcsin	dim	\dim	log	\log
arctan	\arctan	exp	\exp	max	\max
arg	\arg	gcd	\gcd	min	\min
cos	\cos	hom	\hom	Pr	\Pr
cosh	\cosh	inf	\inf	sec	\sec
cot	\cot	ker	\ker	sin	\sin
coth	\coth	lg	\lg	sinh	\sinh
csc	\csc	lim	\lim	sup	\sup
deg	\deg	lim inf	\liminf	tan	\tan
		lim sup	\limsup	tanh	\tanh

Table C.5. Arrow Symbols

←	\leftarrow	⟺	\Longleftrightarrow
⇐	\Leftarrow	⟼	\longmapsto
→	\rightarrow	↪	\hookrightarrow
⇒	\Rightarrow	⇀	\rightharpoonup
↔	\leftrightarrow	⇁	\rightharpoondown
⇔	\Leftrightarrow	⇝	\leadsto
↦	\mapsto	↑	\uparrow
↩	\hookleftarrow	⇑	\Uparrow
↼	\leftharpoonup	↓	\downarrow
↽	\leftharpoondown	⇓	\Downarrow
⇌	\rightleftharpoons	↕	\updownarrow
⟵	\longleftarrow	⇕	\Updownarrow
⟸	\Longleftarrow	↗	\nearrow
⟶	\longrightarrow	↘	\searrow
⟹	\Longrightarrow	↙	\swarrow
⟷	\longleftrightarrow	↖	\nwarrow

Table C.6. Miscellaneous Symbols

ℵ	\aleph	∇	\nabla	♯	\sharp
ℏ	\hbar	√	\surd	\	\backslash
ı	\imath	⊤	\top	∂	\partial
ȷ	\jmath	⊥	\bot	∞	\infty
ℓ	\ell	‖	|	□	\Box
℘	\wp	∠	\angle	◇	\Diamond
ℜ	\Re	∀	\forall	△	\triangle
ℑ	\Im	∃	\exists	♣	\clubsuit
℧	\mho	¬	\neg	◇	\diamondsuit
′	\prime	♭	\flat	♡	\heartsuit
∅	\emptyset	♮	\natural	♠	\spadesuit

Table C.7. Variable-Sized Symbols

∑	\sum	∩	\bigcap	⊙	\bigodot
∏	\prod	∪	\bigcup	⊗	\bigotimes
∐	\coprod	⊔	\bigsqcup	⊕	\bigoplus
∫	\int	∨	\bigvee	⊎	\biguplus
∮	\oint	∧	\bigwedge		

Table C.8. Delimiters

(	(	)	)	↑	\uparrow
[	[	]	]	↓	\downarrow
{	\{	}	\}	↕	\updownarrow
⌊	\lfloor	⌋	\rfloor	⇑	\Uparrow
⌈	\lceil	⌉	\rceil	⇓	\Downarrow
⟨	\langle	⟩	\rangle	⇕	\Updownarrow
/	/	\	\backslash		
|	|	‖	|		

Table C.9. Math Mode Accents

â	\hat{a}	á	\accute{a}	ā	\bar{a}
ǎ	\check{a}	à	\grave{a}	⃗a	\vec{a}
ă	\breve{a}	ã	\tilde{a}	ȧ	\dot{a}
				ä	\ddot{a}

Custom Font Sizes

Different character sizes and typefaces are categorized by font families. A particular size and typeface is called a font. LaTeX uses characters from the Computer Modern (CM) font family designed by Donald Knuth. Some older versions of LaTeX use a variation of the CM fonts called Almost Modern (AM).

LaTeX's default fonts are Computer Modern roman, boldface, italic, small caps, slanted, sans serif, and typewriter. As described in Section 5.2, LaTeX makes it easy to vary these font sizes. This appendix describes how to vary the size of selected fonts using the \newfont command. This command lets you create font sizes not ordinarily supported by LaTeX's built-in size commands. It also lets you use custom fonts that may be supported on your local TeX and LaTeX distribution, as described in your *Local Guide*.

D.1 The "Newfont" Command

LaTeX's "behind-the-scene" default font command is:

```
\newfont{\rm}{cmr10}
```

This instructs TeX to use the 10-point Computer Modern font as it formats your LaTeX input file. Other standard LaTeX typeface commands (e.g. \it, \bf) are predefined to automatically use this same command. These definitions are contained in a file called lfonts.tex.

You can invoke non-standard fonts by using this same command. This does not require you to modify lfont.tex's definitions. For example (and assuming that your site supports this font), to use the 8-point CM sans serif quotation style font, you would type this command in your input file's preamble, just after the \begin{document} command:

```
\newfont{\ssq}{cmssq8}
```

To invoke this font, type \ssq before the section you wish to appear in 8-point CM sans serif quotation style type. After you're through, type \rm to return to the normal CM roman font.

You can create larger fonts by using a variation of this command. A large CM sans serif quotation style font would be created with this command:

```
\newfont{\bigssq}{cmssq8 scaled\magstep5}
```

Use of the "\ssq' and "bigssq' labels is arbitrary; you can name font definitions whatever makes sense.

The scaled\magstep5 statement needs some explanation. If you omit this statement, LaTeX uses the default font size—in this case, 8 points. LaTeX permits you, however, to scale this font to a larger size. This is done by including the scaled\magstep declaration. Following the word magstep, you can use any whole number between 1 and 5 (or larger if your site has generated fonts to support this command). The base magstep is 0, which carries a scale factor of 1,000. One magstep magnifies the base font size by 1.2 times. Magstep1 therefore yields a scaled factor of 1,200. Magstep2 is 1,440, magstep3 is 1,728, magstep4 is 2,074, and magstep5 is 2,488. Larger magsteps produce larger characters. Your LaTeX site administrator can help you figure out what fonts are available on your system.

D.2   Font Samples

Following is a selection of some of LaTeX's most commonly used fonts displayed in scalable sizes. They also list the resulting size in points for each character. Consult your *Local Guide* for additional fonts supported at your site.

Computer Modern Roman (CMR)

5pt cmr5

abcdefghijklmnopqrstuvwxyz ABCDEFGHIJKLMNOPQRSTUVWXYZ 1234567890

6pt cmr6

abcdefghijklmnopqrstuvwxyz ABCDEFGHIJKLMNOPQRSTUVWXYZ 1234567890

7pt cmr7

abcdefghijklmnopqrstuvwxyz ABCDEFGHIJKLMNOPQRSTUVWXYZ
1234567890

8pt cmr8

abcdefghijklmnopqrstuvwxyz ABCDEFGHIJKLMNOPQRSTUVWXYZ
1234567890

9pt cmr9

abcdefghijklmnopqrstuvwxyz ABCDEFGHIJKLMNOPQRSTUVWXYZ
1234567890

10pt cmr10

abcdefghijklmnopqrstuvwxyz 1234567890
ABCDEFGHIJKLMNOPQRSTUVWXYZ

11pt (cmr10 scaled\magstephalf)

abcdefghijklmnopqrstuvwxyz 1234567890
ABCDEFGHIJKLMNOPQRSTUVWXYZ

12pt cmr12

abcdefghijklmnopqrstuvwxyz 1234567890
ABCDEFGHIJKLMNOPQRSTUVWXYZ

14.4pt (cmr10 scaled\magstep2)

abcdefghijklmnopqrstuvwxyz 1234567890
ABCDEFGHIJKLMNOPQRSTUVWXYZ

17.3pt (cmr10 scaled\magstep3)

abcdefghijklmnopqrstuvwxyz
ABCDEFGHIJKLMNOPQRSTUVWXYZ
1234567890

20.7pt (cmr10 scaled\magstep4)

abcdefghijklmnopqrstuvwxyz
ABCDEFGHIJKLM
NOPQRSTUVWXYZ
1234567890

24.9pt (cmr10 scaled\magstep5)

abcdefghijklmnopqrstuvwxyz
ABCDEFGHIJKLM
NOPQRSTUVWXYZ
1234567890

29.4pt (cmr17 scaled\magstep3)

abcdefghijklm
nopqrstuvwxyz
ABCDEFGHIJKLM
NOPQRSTUVWXYZ
1234567890

35.3pt (cmr17 scaled\magstep4)

abcdefghijklm
nopqrstuvwxyz
ABCDEFGHIJKLM
NOPQRSTUVWXYZ
1234567890

Computer Modern Bold Roman (CMBX)

5pt cmbx5

abcdefghijklmnopqrstuvwxyz ABCDEFGHIJKLMNOPQRSTUVWXYZ 1234567890

6pt cmbx6

abcdefghijklmnopqrstuvwxyz ABCDEFGHIJKLMNOPQRSTUVWXYZ 1234567890

7pt cmbx7

abcdefghijklmnopqrstuvwxyz ABCDEFGHIJKLMNOPQRSTUVWXYZ
1234567890

8pt cmbx8

abcdefghijklmnopqrstuvwxyz ABCDEFGHIJKLMNOPQRSTUVWXYZ
1234567890

9pt cmbx9

abcdefghijklmnopqrstuvwxyz 1234567890
ABCDEFGHIJKLMNOPQRSTUVWXYZ

10pt cmbx10

abcdefghijklmnopqrstuvwxyz 1234567890
ABCDEFGHIJKLMNOPQRSTUVWXYZ

11pt (cmbx10 scaled\magstephalf)

abcdefghijklmnopqrstuvwxyz 1234567890
ABCDEFGHIJKLMNOPQRSTUVWXYZ

12pt cmbx12

abcdefghijklmnopqrstuvwxyz 1234567890
ABCDEFGHIJKLMNOPQRSTUVWXYZ

14.4pt (cmbx10 scaled\magstep2)

abcdefghijklmnopqrstuvwxyz 1234567890
ABCDEFGHIJKLMNOPQRSTUVWXYZ

17.3pt (cmbx10 scaled\magstep3)

abcdefghijklmnopqrstuvwxyz
ABCDEFGHIJKLM
NOPQRSTUVWXYZ
1234567890

20.7pt (cmbx10 scaled\magstep4)

abcdefghijklmnopqrstuvwxyz
ABCDEFGHIJKLM
NOPQRSTUVWXYZ
1234567890

24.9pt (cmbx10 scaled\magstep5)

abcdefghijklm
nopqrstuvwxyz
ABCDEFGHIJKLM
NOPQRSTUVWXYZ
1234567890

29.9pt (cmbx12 scaled\magstep5)

abcdefghijklm
nopqrstuvwxyz
ABCDEFGHIJKLM
NOPQRSTUVWXYZ
1234567890

Computer Modern Slant Roman (CMSL)

8pt cmsl8
abcdefghijklmnopqrstuvwxyz 1234567890
ABCDEFGHIJKLMNOPQRSTUVWXYZ

9pt cmsl9
abcdefghijklmnopqrstuvwxyz 1234567890
ABCDEFGHIJKLMNOPQRSTUVWXYZ

10pt cmsl10
abcdefghijklmnopqrstuvwxyz 1234567890
ABCDEFGHIJKLMNOPQRSTUVWXYZ

11pt (cmsl10 scaled\magstephalf)
abcdefghijklmnopqrstuvwxyz 1234567890
ABCDEFGHIJKLMNOPQRSTUVWXYZ

12pt cmsl12
abcdefghijklmnopqrstuvwxyz 1234567890
ABCDEFGHIJKLMNOPQRSTUVWXYZ

14.4pt (cmsl10 scaled\magstep2)
abcdefghijklmnopqrstuvwxyz 1234567890
ABCDEFGHIJKLMNOPQRSTUVWXYZ

17.3pt (cmsl10 scaled\magstep3)
abcdefghijklmnopqrstuvwxyz
ABCDEFGHIJKLM
NOPQRSTUVWXYZ
1234567890

20.7pt (cmsl10 scaled\magstep4)
abcdefghijklmnopqrstuvwxyz
ABCDEFGHIJKLM
NOPQRSTUVWXYZ
1234567890

24.9pt (cmsl10 scaled\magstep5)

abcdefghijklm
nopqrstuvwxyz
ABCDEFGHIJKLM
NOPQRSTUVWXYZ
1234567890

29.9pt (cmsl12 scaled\magstep5)

abcdefghijklm
nopqrstuvwxyz
ABCDEFGHIJKLM
NOPQRSTUVWXYZ
1234567890

Computer Modern Bold Slanted Roman (CMBXSL)

10pt cmbxsl10
abcdefghijklmnopqrstuvwxyz 1234567890
ABCDEFGHIJKLMNOPQRSTUVWXYZ

11pt (cmbxsl10 scaled\magstephalf)
abcdefghijklmnopqrstuvwxyz 1234567890
ABCDEFGHIJKLMNOPQRSTUVWXYZ

12pt (cmbxsl10 scaled\magstep1)
abcdefghijklmnopqrstuvwxyz 1234567890
ABCDEFGHIJKLMNOPQRSTUVWXYZ

14.4pt (cmbxsl10 scaled\magstep2)
abcdefghijklmnopqrstuvwxyz 1234567890
ABCDEFGHIJKLMNOPQRSTUVWXYZ

17.3pt (cmbxsl10 scaled\magstep3)
abcdefghijklmnopqrstuvwxyz
ABCDEFGHIJKLM
NOPQRSTUVWXYZ
1234567890

20.7pt (cmbxsl10 scaled\magstep4)
abcdefghijklmnopqrstuvwxyz
ABCDEFGHIJKLM
NOPQRSTUVWXYZ
1234567890

24.9pt (cmbxsl10 scaled\magstep5)
abcdefghijklm
nopqrstuvwxyz
ABCDEFGHIJKLM
NOPQRSTUVWXYZ
1234567890

Computer Modern Bold Italics (CMBXTI)

10pt cmbxti10
abcdefghijklmnopqrstuvwxyz 1234567890
ABCDEFGHIJKLMNOPQRSTUVWXYZ

11pt (cmbxti10 scaled\magstephalf)
abcdefghijklmnopqrstuvwxyz 1234567890
ABCDEFGHIJKLMNOPQRSTUVWXYZ

12pt (cmbxti10 scaled\magstep1)
abcdefghijklmnopqrstuvwxyz 1234567890
ABCDEFGHIJKLMNOPQRSTUVWXYZ

14.4pt (cmbxti10 scaled\magstep2)
abcdefghijklmnopqrstuvwxyz 1234567890
ABCDEFGHIJKLMNOPQRSTUVWXYZ

17.3pt (cmbxti10 scaled\magstep3)
abcdefghijklmnopqrstuvwxyz
ABCDEFGHIJKLM
NOPQRSTUVWXYZ
1234567890

20.7pt (cmbxti10 scaled\magstep4)
abcdefghijklmnopqrstuvwxyz
ABCDEFGHIJKLM
NOPQRSTUVWXYZ
1234567890

24.9pt (cmbxti10 scaled\magstep5)
abcdefghijklm
nopqrstuvwxyz
ABCDEFGHIJKLM
NOPQRSTUVWXYZ
1234567890

Computer Modern Italics (CMTI)

7pt cmti8

abcdefghijklmnopqrstuvwxyz 1234567890
ABCDEFGHIJKLMNOPQRSTUVWXYZ

8pt cmti8

abcdefghijklmnopqrstuvwxyz 1234567890
ABCDEFGHIJKLMNOPQRSTUVWXYZ

9pt cmti9

abcdefghijklmnopqrstuvwxyz 1234567890
ABCDEFGHIJKLMNOPQRSTUVWXYZ

10pt cmti10

abcdefghijklmnopqrstuvwxyz 1234567890
ABCDEFGHIJKLMNOPQRSTUVWXYZ

11pt (cmti10 scaled\magstephalf)

abcdefghijklmnopqrstuvwxyz 1234567890
ABCDEFGHIJKLMNOPQRSTUVWXYZ

12pt cmti12

abcdefghijklmnopqrstuvwxyz 1234567890
ABCDEFGHIJKLMNOPQRSTUVWXYZ

14.4pt (cmti10 scaled\magstep2)

abcdefghijklmnopqrstuvwxyz 1234567890
ABCDEFGHIJKLMNOPQRSTUVWXYZ

17.3pt (cmti10 scaled\magstep3)

abcdefghijklmnopqrstuvwxyz
ABCDEFGHIJKLM
NOPQRSTUVWXYZ
1234567890

20.7pt (cmti10 scaled\magstep4)

abcdefghijklmnopqrstuvwxyz
ABCDEFGHIJKLM
NOPQRSTUVWXYZ
1234567890

24.9pt (cmti10 scaled\magstep5)

*abcdefghijklm
nopqrstuvwxyz
ABCDEFGHIJKLM
NOPQRSTUVWXYZ
1234567890*

29.9pt (cmti12 scaled\magstep5)

*abcdefghijklm
nopqrstuvwxyz
ABCDEFGHIJKLM
NOPQRSTUVWXYZ
1234567890*

Computer Modern Sans Serif (CMSS)

8pt cmss8

abcdefghijklmnopqrstuvwxyz ABCDEFGHIJKLMNOPQRSTUVWXYZ 1234567890

9pt cmss9

abcdefghijklmnopqrstuvwxyz 1234567890
ABCDEFGHIJKLMNOPQRSTUVWXYZ

10pt cmss10

abcdefghijklmnopqrstuvwxyz 1234567890
ABCDEFGHIJKLMNOPQRSTUVWXYZ

11pt (cmss10 scaled\magstephalf)

abcdefghijklmnopqrstuvwxyz 1234567890
ABCDEFGHIJKLMNOPQRSTUVWXYZ

12pt cmss12

abcdefghijklmnopqrstuvwxyz 1234567890
ABCDEFGHIJKLMNOPQRSTUVWXYZ

14.4pt (cmss10 scaled\magstep2)

abcdefghijklmnopqrstuvwxyz 1234567890
ABCDEFGHIJKLMNOPQRSTUVWXYZ

17.3pt (cmss10 scaled\magstep3)

abcdefghijklmnopqrstuvwxyz
ABCDEFGHIJKLM
NOPQRSTUVWXYZ
1234567890

20.7pt (cmss10 scaled\magstep4)

abcdefghijklmnopqrstuvwxyz
ABCDEFGHIJKLM
NOPQRSTUVWXYZ
1234567890

24.9pt (cmss10 scaled\magstep5)

abcdefghijklm
nopqrstuvwxyz
ABCDEFGHIJKLM
NOPQRSTUVWXYZ
1234567890

29.4pt (cmss17 scaled\magstep3)

abcdefghijklm
nopqrstuvwxyz
ABCDEFGHIJKLM
NOPQRSTUVWXYZ
1234567890

35.3pt (cmss17 scaled\magstep4)

abcdefghijklm
nopqrstuvwxyz
ABCDEFGHIJKLM
NOPQRSTUVWXYZ
1234567890

42.3pt (cmss17 scaled\magstep5)

abcdefghijklm
nopqrstuvwxyz
ABCDEFGHIJKL
MNOPQRSTUVW
XYZ1234567890

Computer Modern Slanted Sans Serif CMSSI)

8pt cmssi8
abcdefghijklmnopqrstuvwxyz ABCDEFGHIJKLMNOPQRSTUVWXYZ 1234567890

9pt cmssi9
abcdefghijklmnopqrstuvwxyz 1234567890
ABCDEFGHIJKLMNOPQRSTUVWXYZ

10pt cmssi10
abcdefghijklmnopqrstuvwxyz 1234567890
ABCDEFGHIJKLMNOPQRSTUVWXYZ

11pt (cmssi10 scaled\magstephalf)
abcdefghijklmnopqrstuvwxyz 1234567890
ABCDEFGHIJKLMNOPQRSTUVWXYZ

12pt cmssi12
abcdefghijklmnopqrstuvwxyz 1234567890
ABCDEFGHIJKLMNOPQRSTUVWXYZ

14.4pt (cmssi10 scaled\magstep2)
abcdefghijklmnopqrstuvwxyz 1234567890
ABCDEFGHIJKLMNOPQRSTUVWXYZ

17.3pt (cmssi10 scaled\magstep3)
abcdefghijklmnopqrstuvwxyz
ABCDEFGHIJKLM
NOPQRSTUVWXYZ
1234567890

20.7pt (cmssi10 scaled\magstep4)
abcdefghijklmnopqrstuvwxyz
ABCDEFGHIJKLM
NOPQRSTUVWXYZ
1234567890

24.9pt (cmssi10 scaled\magstep5)
abcdefghijklm
nopqrstuvwxyz
ABCDEFGHIJKLM
NOPQRSTUVWXYZ
123456789

29.4pt (cmssi17 scaled\magstep3)

abcdefghijklm
nopqrstuvwxyz
ABCDEFGHIJKLM
NOPQRSTUVWXYZ
1234567890

35.3pt (cmssi17 scaled\magstep4)

abcdefghijklm
nopqrstuvwxyz
ABCDEFGHIJKLM
NOPQRSTUVWXYZ
1234567890

42.3pt (cmssi17 scaled\magstep5)

abcdefghijklm
nopqrstuvwxyz
ABCDEFGHIJKL
MNOPQRSTUVW
XYZ1234567890

Computer Modern Bold Sans Serif (CMSSBX)

10pt cmssbx10
abcdefghijklmnopqrstuvwxyz 1234567890
ABCDEFGHIJKLMNOPQRSTUVWXYZ

11pt (cmssbx10 scaled\magstephalf)
abcdefghijklmnopqrstuvwxyz 1234567890
ABCDEFGHIJKLMNOPQRSTUVWXYZ

12pt (cmssbx10 scaled\magstep1)
abcdefghijklmnopqrstuvwxyz 1234567890
ABCDEFGHIJKLMNOPQRSTUVWXYZ

14.4pt (cmssbx10 scaled\magstep2)
abcdefghijklmnopqrstuvwxyz 1234567890
ABCDEFGHIJKLMNOPQRSTUVWXYZ

17.3pt (cmssbx10 scaled\magstep3)
abcdefghijklmnopqrstuvwxyz
ABCDEFGHIJKLM
NOPQRSTUVWXYZ
1234567890

20.7pt (cmssbx10 scaled\magstep4)
abcdefghijklmnopqrstuvwxyz
ABCDEFGHIJKLM
NOPQRSTUVWXYZ
1234567890

24.9pt (cmssbx10 scaled\magstep5)
abcdefghijklm
nopqrstuvwxyz
ABCDEFGHIJKLM
NOPQRSTUVWXYZ
1234567890

Computer Modern Caps and Small Caps (CMCSC)

10pt cmcsc10
ABCDEFGHIJKLMNOPQRSTUVWXYZ 1234567890
ABCDEFGHIJKLMNOPQRSTUVWXYZ

11pt (cmcsc10 scaled\magstephalf)
ABCDEFGHIJKLMNOPQRSTUVWXYZ 1234567890
ABCDEFGHIJKLMNOPQRSTUVWXYZ

12pt (cmcsc10 scaled\magstep1)
ABCDEFGHIJKLMNOPQRSTUVWXYZ 1234567890
ABCDEFGHIJKLMNOPQRSTUVWXYZ

14.4pt (cmcsc10 scaled\magstep2)
ABCDEFGHIJKLMNOPQRSTUVWXYZ 1234567890
ABCDEFGHIJKLMNOPQRSTUVWXYZ

17.3pt (cmcsc10 scaled\magstep3)
ABCDEFGHIJKLMNOPQRSTUVWXYZ
ABCDEFGHIJKLM
NOPQRSTUVWXYZ
1234567890

20.7pt (cmcsc10 scaled\magstep4)
ABCDEFGHIJKLMNOPQRSTUVWXYZ
ABCDEFGHIJKLM
NOPQRSTUVWXYZ
1234567890

24.9pt (cmcsc10 scaled\magstep5)
ABCDEFGHIJKLM
NOPQRSTUVWXYZ
ABCDEFGHIJKLM
NOPQRSTUVWXYZ
1234567890

Computer Modern Inch High Sans Serif (CMINCH)

cminch

ABCD
EFGH
IJKL
MNO
PQRS

TUV
WXY
Z123
4567
890

Computer Modern Typewriter (CMTT)

9pt cmtt9
abcdefghijklmnopqrstuvwxyz ABCDEFGHIJKLMNOPQRSTUVWXYZ 1234567890

10pt cmtt10
abcdefghijklmnopqrstuvwxyz 1234567890
ABCDEFGHIJKLMNOPQRSTUVWXYZ

11pt (cmtt10 scaled\magstephalf)
abcdefghijklmnopqrstuvwxyz 1234567890
ABCDEFGHIJKLMNOPQRSTUVWXYZ

12pt cmtt12
abcdefghijklmnopqrstuvwxyz 1234567890
ABCDEFGHIJKLMNOPQRSTUVWXYZ

14.4pt (cmtt10 scaled\magstep2)
abcdefghijklmnopqrstuvwxyz 1234567890
ABCDEFGHIJKLMNOPQRSTUVWXYZ

17.3pt (cmtt10 scaled\magstep3)
abcdefghijklmnopqrstuvwxyz
ABCDEFGHIJKLM
NOPQRSTUVWXYZ
1234567890

20.7pt (cmtt10 scaled\magstep4)
abcdefghijklmnopqrstuvwxyz
ABCDEFGHIJKLM
NOPQRSTUVWXYZ
1234567890

24.9pt (cmtt10 scaled\magstep5)

abcdefghijklm
nopqrstuvwxyz
ABCDEFGHIJKLM
NOPQRSTUVWXYZ
1234567890

Glossary

Note: words in SMALL CAPS are defined elsewhere in the Glossary.

Alignment. Manner in which text lines up in a column or page. Alignment can be flush against the left margin, flush against the right margin, justified flush against both left and right margins, and centered (no LEFT or RIGHT JUSTIFICATION).

Array. Matrix of mathematical elements arranged in columns and rows.

Article Style. LaTeX DOCUMENT STYLE used to produce articles such as submissions to scholarly journals.

ASCII File. Acronym for *American Standard Code for Information Interchange.* Pure text file containing no special formatting symbols. LaTeX INPUT FILES must be in this format.

Base Line. In type, an imaginary line connecting the bottoms of all capital letters on a line.

Baselineskip. LaTeX command to adjust the vertical distance between the bottom of a BASE LINE, and the bottom of the next base line in the same paragraph.

Baselinestretch. LaTeX command consisting of a decimal number (e.g. 1.5 or 2); default value is 1. Used to change BASELINESKIP to default value times baselinestretch, often to create a double-spaced effect. Changed with LaTeX RENEWCOMMAND.

Bezier Style. LaTeX command used to draw a curved line in a PICTURE environment.

Bitmapped Font. Each character SIZE and STYLE is designed and stored in separate FONT files. LaTeX and TeX use bitmapped fonts to generate output.

Body. Main part of document's text, excluding HEADER, FOOTER, and SECTIONAL headings.

Bold Face. Heavy, thicker line version of standard TYPEFACE.

Book Style. LaTeX DOCUMENT STYLE used to produce books.

Box. Block of text treated by TeX as a single unit; can be as small as a single character. LaTeX makes three kinds of boxes: LR boxes which process text in LR MODE (left-to-right); PARBOXES which are text processed in paragraph mode; and rule boxes which are a rectangular blotch of ink.

Camera Copy, Camera Ready. Copy prepared for photoreproduction; requires no further editing or changes.

Caps, Small Caps. Caps are text in all upper-case letters; small caps are capital letters in the same size as lower-case letters for the same font.

Center Environment. LaTeX environment that centers all text. Usually leaves multiline text ragged on both left and right.

Columnsep. LaTeX command to adjust space width between text columns in the TWOCOLUMN STYLE.

Columnseprule. LaTeX command to adjust width of vertical line separating text columns in the TWOCOLUMN STYLE. Default value is zero (an invisible line).

Compositor. Person who sets a manuscript into type.

Cross-Reference. Notation at point in document to relevant information located at another place in the document. LaTeX permits cross-references to both forward and reverse locations.

Delimiter. Symbol used by LaTeX to mark the beginning and end of a command phrase or ENVIRONMENT.

Descriptive Markup System. System like LaTeX where a COMPOSITOR inserts TYPESETTING commands into a document file for processing by a FORMATTING PROGRAM. Commands describe document elements such as quotation, list, table, and chapter. Final output is controlled by predefined DOCUMENT STYLES.

Displaymath Environment. LaTeX environment which centers mathematical equations, formulas, and expressions in the center of a page following normal text.

Document Style. Predetermined style that determines layout for entire document. Standard LaTeX document styles include ARTICLE, BOOK, LETTER, and REPORT.

DVI File. Acronym for *device-independent* file. A DVI file results from processing an INPUT FILE through TeX or LaTeX. DVI files can be processed by a PRINTER DRIVER for output on a printer, or previewed on a display monitor.

Em Quad or Em. In printing, unit of linear measurement equal to POINT size of FONT in use. Usually equal to the width of a capital M.

Em-Dash. Wide dash used to connect compound phrases. In LaTeX, created by typing three hypens (e.g. ---). Longer than an EN-DASH.

Emphasized. Emphatic LaTeX TYPEFACE. If text is roman, emphatic typeface is ITALIC. If text is italic, emphatic typeface is ROMAN.

En Quad or En. In printing, one half of an EM QUAD.

En-Dash. Wide dash used to connect number ranges. In LaTeX, created by typing two hyphens (e.g. --). Shorter than an EM-DASH.

Environment. Self-contained section of text upon which a special operation is performed. An ITALIC environment, for example, is a section of text that prints in an italic font. Environments can vary in size between a single character and an entire document.

Eqnarray Environment. Multiline equation environment that functions like a three-column math ARRAY. Does not work in MATH mode because it creates its own math mode automatically.

Equation Environment. Like DISPLAYMATH except equation number is automatically added against margin. This number automatically increases itself in each BOOK chapter or ARTICLE.

Error Message. Messages displayed while running LaTeX that alert COM-POSITOR to errors contained in INPUT FILE. Errors generally result from mistyped commands or misused commands.

Evensidemargin. LaTeX command to define amount of left MARGIN on even-numbered, or left-handed pages, in documents formatted with the TWOSIDE document style option.

Figure Environment. Allows tables, text, or a predefined amount of blank space to be labeled and positioned to an optimal spot in a document. Similar to TABLE environment.

Fill Environment. Makes text expand as much as it can in a given space. Default distance is zero, yet fill is infinitely stretchable. Can be assigned finite values to produce rigid lengths of SPACE in text.

Fleqn Environment. DOCUMENT STYLE option to cause math equations and expressions in DISPLAYMATH and EQUATION environments to be displayed flush against left MARGIN.

Flush Left. Text lines are vertically aligned at same point from left side of page, and can end at varying points from right side of page. Opposite of FLUSH RIGHT.

Flush Right. Text lines can begin at varying points from left side of page, but are vertically aligned at the same point from right side of page. Opposite of FLUSH LEFT.

Flushleft Environment. Creates FLUSH-LEFT text.

Flushright Environment. Creates FLUSH-RIGHT text.

Font. Family or set of characters in same TYPEFACE and SIZE. Characters include letters, punctuation marks, and symbols. For example, Computer Modern roman 10-point font differs from Computer Modern sans serif 10-point font; it also differs from Computer Modern roman 12-point font.

Footer. Recurring text at bottom of each page.

Footheight. LaTeX command to set height of BOX containing FOOTER.

Footskip. LaTeX command to set distance between BASELINE of a page's last line of text and the FOOTER's baseline.

Formatting Program. Software that processes an INPUT FILE into final typeset form. TEX is the formatter for LATEX input files.

Front Matter. Material such as title page, preface, and table of contents that precedes text.

Galley, Galley Proof. TYPESET sheet of copy, often in long strips of coated paper. Usually given to author for proofreading.

Glue. Concept to describe "substance" that acts as an adhesive between BOXES. Glue has three characteristics manipulated with TEX commands: it consists of space, it can stretch, and it can contract.

Gutter. Inner margins of facing pages in a book.

Header. Recurring text at top of each page.

Headheight. LATEX command to set the distance between the bottom and top of a HEADER.

Headsep. LATEX command to set the distance between the HEADER BASE LINE and top of the first line of text.

Helvetica. Special type of SANS SERIF FONT.

Hyphen. Horizontal line between two characters used to divide a word between two lines.

Hyphenation. Process of determining where words should break between two lines with a HYPHEN.

Ifthen Style. STYLE option that allows use of two Pascal language structures in INPUT FILES: (1) if, then, else and (2) while, do arguments.

Input File. Source ASCII TEXT FILE processed by LATEX. Contains a document's text and MARKUP format commands.

Italic Correction. Technique used to add a little space between an ITALIC character and a normal character that follows. Typically done with \/ or {} character pairs.

Italic. TYPEFACE that slants to the right.

Justified Text. Text that is both FLUSH LEFT and FLUSH RIGHT.

Kerning. Spacing between two letters. Formulas that depend on letter combinations determine amount of spacing. For example, the AW combination has more space between its letters than does MN.

Key Field. The phrase typed in the text body that ties citations into a bibliographic database.

LaTeX. DESCRIPTIVE MARKUP document processing system created by Leslie Lamport. Collection of special commands, or MACROS, based on TeX typesetting system created by Donald E. Knuth.

Leading. Measure of vertical distance between two lines of text. POINT measurement usually refers to distance between two base lines.

Letter Style. DOCUMENT STYLE used to format letters. Contains commands to produce elements such as the date, address, opening, closing, and signature.

Ligature. Letter combinations joined together as one unit such as ff, fi, fl, ffi, and ffl.

Local Guide. Booklet written and distributed by a local TeX and LaTeX administrator that describes how to install LaTeX, process an INPUT FILE, use locally owned printers, and take advantage of features implemented by local technical staff (e.g. company logo on letterhead).

Log File. Text file in which ERROR MESSAGES are recorded.

LR Mode. Left-to-right mode consists of LaTeX commands that instruct TeX to format text as a series of words separated by spaces. TeX does not create new typeset lines while processing LR mode. Can be nested within PARAGRAPH or MATH MODES.

Macro. Technique where one command substitutes as a shortcut to execute several commands. LaTeX is a macro implementation of TeX because it compresses many TeX PROCEDURAL commands into a few LaTeX DESCRIPTIVE commands.

Magstep. TeX command used to increase a FONT's default size in incremental steps.

Margin. White space that surrounds text BODY on left and right, HEADER on top, and FOOTER on bottom of page.

Marginparpush. LaTeX command to set minimum vertical distance between two marginal notes.

Marginparsep. LaTeX command to set distance between a margin and marginal note.

Marginparwidth. LaTeX command to set width of a marginal note.

Markup System. A formatting scheme like LaTeX by which ASCII commands in a TEXT FILE provide formatting instructions to a program through which the file is processed, after text is typed with a text editor or word processor.

Math Mode. ENVIRONMENT that creates math formulas, expressions, and symbols on a line of normal text. Can be nested within PARAGRAPH or LR MODES.

Matrix. Mathematical elements in a rectangular arrangement of columns and rows.

Measurement. Valid units of measurement for LaTeX include inches, millimeters, centimeters, as well as the printer's EM, EX, PICA, and POINT.

Minipage Environment. Allows placement of a PARBOX of text at a particular spot. One application is to create parallel BOXES of text without using the TWOCOLUMN STYLE option.

Negate Sign. In math, denotes a negative number. Its HYPHEN is slightly closer to the number or symbol than a minus sign.

Newcommand. LaTeX command to create a MACRO.

Newenvironment. LaTeX command to create a new ENVIRONMENT.

Newfont. LaTeX command to define a new FONT other than the default Computer Modern roman font. New font definitions usually go in the INPUT FILE's PREAMBLE. Once defined, new fonts are usable upon demand.

Newpage. LaTeX command to force a page break. When used in the TWOCOLUMN DOCUMENT STYLE option, this command forces text to the next column.

Oddsidemargin. LaTeX command to define amount of left MARGIN on odd-numbered, or right-handed pages, in documents formatted with the TWOSIDE DOCUMENT style option.

Orphan. First line of a paragraph that is isolated at the bottom of a page or column. Similar to a WIDOW.

Outline Font. Each character is designed strictly to define its shape, and stored as a curve-fitting equation. Printers that use a programming language such as POSTSCRIPT, use this equation to produce the TYPEFACE in any SIZE.

Parbox. BOX typeset in ordinary PARAGRAPH MODE. Created automatically by TABLE and FIGURE environments. Created manually with LaTeX parbox and MINIPAGE environments.

Paragraph Mode. TeX's normal method of processing words and sentences into lines, paragraphs, and pages.

Part. LaTeX sectional unit command that includes one or more chapters.

Pica. Printer's unit of measurement where one pica equals 12 POINTS.

Picture Environment. LaTeX environment used to draw pictures of circles, straight lines, arrows, and text. Objects are positioned with x and y coordinate specifications.

Point. Printer's unit of measurement where one inch equals 72.27 points.

PostScript. Printer control language licensed by Adobe Systems. PostScript translates OUTLINE FONTS into print on laser printers that use its language.

Preamble. Collection of LaTeX commands typed at beginning of INPUT FILE before the \begin{document} command. Contains setup commands such as NEWFONT, NEWCOMMAND, and page layout redefinitions.

Printer Driver. Program to convert a DVI file into a form capable of being printed on a specific output device.

Procedural Markup System. MARKUP SYSTEM like TeX in which virtually every formatting procedure must be explicitly stated by commands (e.g. margins, spacing between lines and paragraphs, character sizes, etc.).

Proceedings Style. ARTICLE DOCUMENT STYLE option that creates two-column output tailored for ACM and IEEE conference proceedings.

Proportional Spacing. Spacing between characters is based upon each character's particular width. Generally, narrow characters have less separation space than wider characters.

Punctuational Markup System. MARKUP SYSTEM wherein all formatting is manually typed by COMPOSITOR. Punctuation elements like periods, commas, and quote marks are found in all types of markup systems, word processors, and text editors.

Quotation Environment. Indents text on both left and right sides from normal left and right margins. Also indents first line of each paragraph.

Quote Environment. Indents text on both left and right sides from normal left and right margins. Does not indent first line of each paragraph.

Ragged Text. Text that has an uneven left or right margin. Centered blocks of text usually have ragged left and right margins.

Report Style. DOCUMENT STYLE used for technical reports.

Roman. Type that is vertically straight, or unslanted.

Running Head. A HEADER contained on each page. Often includes chapter or section number, chapter or section name, or author name(s).

Sans Serif. TYPEFACE that does not include SERIFS.

Sectioning. Divisional units in an article, report, or book.

Serifs. Small lines or curves that dress up the ends of a character's main strokes.

Size. Vertical dimension of character FONTS.

Slanted. Type that is slanted, like ITALIC type.

Small Caps. See CAPS.

Space. Distance or gap between two points on a page.

Spacing. Describes lateral distancing between characters. See EM QUAD or EM.

Stacked Fraction. In math, a fraction where the numerator is set above a ruled line and the denominator is set below the same line.

Style File. LaTeX file used to control the DOCUMENT STYLE.

Style. Rules that uniformly govern page dimension, character SPACING and LEADING, FONT types and SIZES, etc.

Subscript. Character(s) or symbol(s) that print below a BASE LINE.

Superscript. Character(s) or symbol(s) that print above a BASE LINE such as in a footnote reference.

Tabbing Environment. Allows creation of tabs, or systematic indentations of text.

Table Environment. Allows tables, text, or a predefined amount of blank space to be labeled and positioned to an optimal spot in a document. Similar to FIGURE environment.

Tabular Environment. Allows tables, text, or a predefined amount of blank space to be labeled and positioned to an optimal spot in a document. Does not allow labeling of this material as the TABLE ENVIRONMENT does.

TeX. PROCEDURAL MARKUP TYPESETTING system created by Donald E. Knuth, written in Web programming language.

Text File. A computer file that consists solely of ASCII characters.

Textheight. LaTeX command to set the height of the BODY on each page.

Textwidth. LaTeX command to set the width of the BODY on each page.

Tie. Synonymous to use of a ˜ (tilde) character that causes TeX to create an ordinary interword space without breaking the line at that point.

Times Roman. ROMAN TYPEFACE created in 1931 for the London *Times*. One of the most popular typefaces used in TYPESETTING.

Topmargin. LaTeX command to set distance from a point one inch from top of page to the top of a page HEADER.

Twocolumn Style. DOCUMENT STYLE option to create two columns of text on each page.

Twoside. DOCUMENT STYLE option that formats document for printing on both sides of each page. This is the default setting for the BOOK document style.

Type Size. Physical size of a type FONT, usually vertically measured in POINTS.

Typeface. The style of type. Examples include BOLDFACE, ITALIC, ROMAN, SANS SERIF, and TYPEWRITER.

Typesetting. The act of setting a manuscript into type.

Typewriter. TYPEFACE that appears created on a typewriter.

Unary Operator. In math, a negative number or character.

Underline. A ruled line that appears underneath a character or string of characters.

Verb Environment. VERBATIM ENVIRONMENT created on a single line.

Verbatim Environment. Reproduces text exactly as it appears in the INPUT FILE. Output appears in TYPEWRITER TYPEFACE.

Verse Environment. Creates output in form usually reserved for poetry.

Visual Markup System. All formatting is manually typed by compositor. Punctuational elements like periods, commas, quote marks, and space between paragraphs visually format the document.

Widow. Last line of a paragraph that is isolated at the top of a new page or column. Similar to an ORPHAN.

Bibliography

[1] *America—20th Century Poetry: Landscapes of the Mind.* McDougal, Littell & Co., Evanston, Ill., 1973.

[2] *The Chicago Manual of Style.* The Univ. of Chicago Press, 13th edition, 1982.

[3] James H. Coombs, Allen H. Renear, and Steven J. DeRose. "Markup Systems and the Future of Scholarly Text Processing". *Communications of the ACM*, 30(11):933–47, 1987.

[4] Donald E. Knuth. *The TEXbook.* Addison-Wesley Publishing Co., Reading, Ma., 1986.

[5] Leslie Lamport. "Document Production: Visual or Logical?". *Notices of the American Mathematical Society*, 34(4):621–24, 1987.

[6] Leslie Lamport. *LATEX: A Document Preparation System.* Addison-Wesley Publishing Co., Reading, Ma., 1986.

[7] Edward Connery Lathem, editor. *The Poetry of Robert Frost.* Holt, Rinehart and Winston, New York, 1969.

[8] Richard Palais. "Mathematical Text Processing". *Notices of the American Mathematical Society*, 35(3):391–96, March 1988.

[9] I. S. Shklovskii and Carl Sagan. *Intelligent Life in the Universe.* Dell Publishing Co., Inc., New York, 1966.

[10] Sina Spiker. *Indexing Your Book: A Practical Guide for Authors.* Univ. of Wisc. Press, Madison, Wisc., 1954.

[11] William Strunk, Jr. and E. B. White. *The Elements of Style.* The Macmillan Co., New York, 1972.

[12] Ellen Swanson. *Mathematics Into Type: Copy Editing and Proofreading of Mathematics for Editorial Assistants and Authors.* American Mathematical Society, Providence, RI, 1986 reprint edition.

[13] Mary-Claire van Leunen. *A Handbook for Scholars.* Alfred A. Knopf, New York, 1979.

Index

About the Author

David J. Buerger is Executive Editor/Connectivity & Testing with *Infoworld*, a leading news and information weekly for computer professionals. Previously he was Director of the Personal Computer Center at Santa Clara University, where he spearheaded the introduction and widespread use of LaTeX by scientists and engineers.